Last Ranch in Hells Canyon

Further Adventures of the Mantle Family

By Queeda Mantle Walker

LIFETIME CHRONICLE PRESS

Montrose, CO

© 2009 Queeda Mantle Walker
All rights reserved in whole or in part.

First edition
Printed in the United States of America

Library of Congress Control Number: 2008942333

ISBN: 978-0-9814539-6-5

Cover photo from the Kail Mantle collection by
Will Brewster Photography
Back cover photo from Classic Photography by Cheryl Haines

Copies of the book may be ordered through:
Museum of Northwest Colorado
970-824-6360
Fax: 970-824-1098
www.museumnwco.org
musnwco@moffatcounty.net

Published by:
Lifetime Chronicle Press
121 N. Park Ave.
Montrose, CO 81401
970-240-1345
chronicle@montrose.net

Table of Contents

INTRODUCTION. 1
CHAPTER 1: What Now?. 3
   Charley and Evelyn: Life in Brief, Fall 1949 to 1965
CHAPTER 2: Memories, 1958 10
CHAPTER 3: Charley, Jr. 14
CHAPTER 4: Lonnie (Nav), Queeda, & Pat Mantle, 1953 to 1959 . 17
CHAPTER 5: Charley Goes For It, 1960 20
CHAPTER 6: Corumba 25
CHAPTER 7: The State of Goias 28
CHAPTER 8: Cowboy in the Jungle 32
CHAPTER 9: Charley Gets It Done 36
CHAPTER 10: Charley Up the Amazon 40
CHAPTER 11: Charley Up the Pan-American Highway. 43
CHAPTER 12: Evelyn, 1959 to 1961 47
CHAPTER 13: Wild Horses. 53
CHAPTER 14: Lonnie Moves to Wyoming. 60
CHAPTER 15: Australia, 1962 64
CHAPTER 16: Never Say Retire, 1962 67
CHAPTER 17: Charley, Jr., Fall 1962 to 1965 80
CHAPTER 18: The Mantle Ranch, 1964 to 1965 85
CHAPTER 19: Mantle Ranch Dilemma, 1965 90
CHAPTER 20: Lonnie's Shoeing Chute, 1965 to 1966. 96
CHAPTER 21: Marmion to Colorado, 1966 99
CHAPTER 22: Grandpa in Montana – 1966 105
CHAPTER 23: Evelyn to Boulder, 1966 to 1967 107
CHAPTER 24: Browns Park, Jay Road, and Charley, 1967 to 1969 115
CHAPTER 25: Charley, Jr.—Potch, 1970 to 1971 125
CHAPTER 26: Lonnie to Pavillion, 1973. 128
CHAPTER 27: Evelyn, 1970 to 1979 134
CHAPTER 28: Pat, 1967 to 1992 142
CHAPTER 29: Mantle Ranch vs. United States Park Service,
   et al 151
EPILOGUE. 170
CHRONOLOGICAL DATES OF EVENTS 172

Introduction

IN 1926, CHARLEY MANTLE AND EVELYN FULLER were married. Life led them to live in one of the most difficult places in America—the Yampa River Canyon. Without roads, electricity, running water, or telephones, they were self-sufficient and taught their children to be also.

As this story tells, they created a dynasty of Western ranchers. Their children have grown up, dispersed, and are carrying on the knowledge and traditions of ranch life that their parents taught them so well. Charley and Evelyn were proud to see that their children were able to compete and succeed in the modern world.

To learn more about this startling couple, read the book *The Mantle Ranch, A Family's Joys and Sorrows in the Beautiful, Remote Yampa River Canyon* by Queeda Mantle Walker, ISBN 0-87108-350-7.

1

What Now?

Charley and Evelyn: Life in Brief, Fall 1949 to 1965

THE MANTLE RANCH WAS IN for some big changes this year. It had unmatched scenery, nice big house, a healthy herd of Herefords, good saddle horses—all a cowboy's dream. But Charley Mantle's youngest child would be going away this fall to boarding school. That would mean he and Evelyn, his wife, would be all alone.

Come September, Tim would go to Wasatch Academy in Mt. Pleasant, Utah, as an eighth grader, and Lonnie would be a junior there. Queeda was now a sophomore in college at CU-Boulder, Pat had a job at the neighbor's ranch, and God only knew where their oldest son, Charley, Jr.— "Potch" was. That left Charley and Evelyn all alone at the ranch. The nearest neighbor was forty miles away now that the Chew family, who used to live ten miles away, had moved out. That had never bothered Charley before, but age and his work-abused body were not like they used to be. He guessed they would just do the best they could, like they always did.

After delivering Queeda to Boulder, which was three hundred mile away, they returned to the ranch, and grim reality. Charley had intense back pain from an old injury when his horse had thrown him in a rock pile some years ago. All of his fifty-eight years of wrenched bones and overworked ligaments ached as well. He was unable to do hardly any work. Evelyn was overworked trying to carry the load for both of them.

Charley laid on the cot in the living room, struggling to learn Spanish. He dreamed of going to Brasil. His mission was to find ranch country where his boys could have a big ranch with no government interference like he had endured all his life. He studied the Spanish dictionary and learned many words—enough to get by—but meantime his hearing was getting worse. Everything anyone said to him had to be repeated in a very loud shout. It didn't take long for people to get tired of that. He started thinking seriously about getting a hearing aid.

Pat begged Charley and Evelyn to pack up the truck and go to Mexico for the winter. He said he could look after the ranch—and please, would they go have a vacation and get healed. They all knew that rest was the only thing that was going to help Charley's poor back. Evelyn badly needed the rest, too.

They would not leave Pat by himself, but fate had different plans. When bringing their children home for Christmas, the whole family nearly perished on the snow-choked top of Blue Mountain. Charley was terribly sick, and when word could finally be gotten out, he was hauled out in a snowplow to the doctor in Craig. A Moffat County road crew under supervisor Dick Toole's guidance had come after him. Dick was a longtime friend of the Mantles. Dr. Monahan told Charley that he had a bad case of pneumonia and would die if he went back down to the ranch to work in bitter cold weather.

They stayed with Dick and Maude Toole in Craig while Charley healed. Dick helped them build a chuck wagon that fit snugly in the back of the pickup. They threw in their bedroll and a mess kit and took off for a camping trip in Mexico. They traveled in Chihuahua and Sinaloa states. They enjoyed every minute, and were welcomed by the Mexicans they met. Topolobampo and Los Mochis in Sinaloa were their favorite places. They rested up and came home in the early spring.

It had been the cruelest winter in history. It wiped out livestock herds all through the West. It had killed most of the Mantle cattle herd and nearly all the horse herd. Pat had saved those he could reach, but it was way more than any human could do to save all of them. Pat barely made it through himself. His eyes were dark holes of pain and visions of too much death.

When the snow melted in the spring, devastating floods roared down Hells Canyon, destroying dams, fields, and hay crops, and left the ranch once more without a road. Their only neighbors, the Chew family, never moved back to the ranch. They sold out to the National Park Service.

To add to Evelyn's misery, her mother, Julia Fuller, died, and Evelyn was unable to attend the funeral. Charley picked up the telegram in Youghal informing Evelyn of her mother's death, but he failed to deliver the telegram promptly so Evelyn could attend her mother's funeral. Another serious problem for Evelyn was that the Jeep tires were patches on patches, and Charley wouldn't take it in and have it fixed, so they were stranded there in the canyons alone. Even their walkie-talkie, which connected with Dinosaur headquarters, was not working

Charley's arthritis had his left arm nearly paralyzed, so he was able to do very little work. Pat would be drafted into the army soon, so there would be nobody to hold the place together. Evelyn was in a state of deep melancholy by March, 1952. For the first time in her life she had to search her soul for a purpose to her life.

In the spring of 1954, Queeda chose not to come home after she graduated from college. She was desperate to get a job and use her college degree. She knew if she came home, she would never do that. It was another terrible blow to Charley and Evelyn, losing their only daughter.

It looked like the Echo Park Dam, a reclamation project downriver from the Mantle Ranch, where the Yampa River joins the Green River, was a sure thing to happen. This dam would bury the Mantle Ranch under 200 feet of water. The government would condemn their property and buy it for almost nothing, stripping them of their livelihood. Mantles contacted their representatives for assistance. As a result of pressure from many of their constituents, politicians became very active in either condemning this proposed dam, or championing it. A great tug of war took place, and this project—the Echo Park Dam—is widely recognized as the beginning of active involvement by conservation groups in Bureau of Reclamation projects in the United States. In this case it was the Sierra Club that came forward and campaigned vigorously against the Echo Park Dam. Some of their members had floated the Yampa River and recognized the canyons as a beautiful, unique American treasure that needed to be preserved. The final result in years to come was that the project was defeated and the Mantle Ranch was not to end up buried under 200 feet of water.

Charley began having blackouts. He would stand up, stagger a little, and fall over. Evelyn assumed it was high blood pressure. She

didn't dare leave him to go get help, and he wouldn't let her take him out to a doctor. He went on a self-managed diet of no fats and lost some weight and did feel somewhat better. These were very difficult years for them both.

※ ※ ※

Come spring of 1955, Queeda married Rex Walker. Rex's first job was with Socony Mobile Oil Company in Venezuela. They went away to live in Caracas for two years, and Evelyn became distraught beyond words. She just knew something terrible would happen to them, and she would never see Queeda again.

※ ※ ※

All these changes to their lives weighed heavily on Charley and Evelyn. The youth and vigor had run out on these two aging warriors. They knew that Tim and Lonnie needed to have their own lives, but what with serving their time in the army and taking turns tending the old ranch, the boys felt trapped.

February, 1956, Evelyn wrote to her cousin, Eva:

Our neighbor is talking of going out to Vernal by horseback the 20th, so I'll get some mail ready to go along. It seems a long time since our last mail.

We are certainly having an unusual winter. Each day for the past two weeks it has been in the high 40s, and yesterday was 52 while Denver broke a record of 69. Our cold and snowy weather was during November and December but was also very cold while Charley and Pat were up-country gathering the stock to bring down to feed. They got back in on Groundhog day. After being gone nearly a week I really got uneasy about them and went over and got Rial [Chews still had some grazing rights, and he spent the winter taking care of his sheep on the range at their old ranch] to help me get a horse so we could go see what was the matter, but they came in that night. It had been slow going, for snow was above the horses' knees up there at Schoonover (east of here – not on mountain top) and the trails into the canyon were icy and hard to make the stock take it.

Have about everything on the feed yard now and driving out to Red Rock each day to feed hay. Since then it has turned so warm that yesterday they went downriver and got calves up out of the canyon before the ice broke up. The washes are running water and will soon raise the river till the ice will begin moving.

I was going out with them each day to feed but got too ambitious and wrenched my back helping load the hay bales and have been wearing a dress and playing <u>only a housewife</u> ever since. The staying inside makes me anxious to be able to throw open the doors and get to house cleaning. The curtains and corners show winter's buildup of cobwebs and soot.

Aunt Alice I want to thank you so much for the mittens. Those are the first hand knits I've had since the last ones you sent several years ago and I appreciate the work that goes into them, for I've tried my hand at knitting enough to know that it sure takes patience and perseverance. Which I'm often running out of before I get the job done.

I'd planned on doing so much this winter in sewing but while the days are so short, just everyday tasks take up most of the daylight. I'll have to get the bug soon though, for I'm completely out of house dresses. The ones I have left won't make good rug rags. Queeda left a lot of clothes when she got married and I've been wearing those out.

Which makes me wonder if you know Rex and Queeda are on their way home via Central America. She wrote a letter to Dinosaur Headquarters at Jensen, from Cristobal, Panama Canal Zone and said they were waiting for a boat to take them through the canal and up to Costa Rica. Pat and Sue had a letter he told us about when he got back home about the eleventh of Jan. and they planned to leave Venezuela on January twentyfifth and be in the States around the first of March. Her letter from Cristobal was dated the fifth of February. Wonder if it took them 10 days by boat to there. I am so anxious to see them I'm wishing away this snow by sheer will power. They are coming to visit as soon as they get back.

Our area has been a hot box of planes of late. A two engine Cessna is down between Ogden and Denver with four men of Denver aboard. They have been searching the area for a solid week but not a trace. Pat found shreds of what looks like foil all over the countryside but it is all the same length and width— something the order of Christmas tinsel only not so wide. About like the small end of a tooth pick and one to two inches long. Since his finding it we haven't been able to contact Dinosaur headquarters on the radio. Doubtful if its connected with the down plane but they are asking for any clue whatever from farmers and ranchers in the flight path. There are 75 planes in the air searching today.

Eva, I've sure been enjoying the books you sent, especially since my back gives me an excuse to lie around.

Charley has kept going pretty good but it would be impossible without Pat and its hard on him too with his family one place and he another. They know it's wiser they wait till Spring to bring the wee one in. I was so darn cock-sure when mine were babies—now I wonder how I got that way, but then nature was kind to me too for I was a veritable milk cow and had no worries over formulas etc, besides being such a healthy clod. I appreciate that more now as my near-fifty years, at times, become more evident so that I must slow down now and then.

Pat and I have been tanning some skins—we have a beautiful calf skin tanned and a huge bobcat in the liquor to set the hair – next is to be a tiny civet cat, a cousin of a skunk, and we hope to tan a fox. The hard work is working them after they come out of the liquor which is five gallons water, two pounds salt and one pound alum, but the calf turned out nice. I hope to get some felt and back it and the others. This is an unborn calf and cuter than anything. Its mom died two years ago and Pat skinned it out then and dried it. He also mounted a set of bull horns that look unreal they are so massive. Belonged to Pete who died a few years back and was best bull we ever owned. Have his name burned on the mounting board.

Best love to you all, Evelyn

Queeda and Rex moved back to Texas, Rex's home state, after he finished his hitch in Venezuela. In December 1957 they had a baby—Cindy Evelyn Walker—born December 12, 1957, in Tyler, Texas. Evelyn traveled to Texas to help out. There she worked too hard tending the mother and baby. Then she overworked helping the family move to Colorado when the baby was only two weeks old. Rex planned to complete his master's degree in Geology at Colorado University.

Evelyn returned to the ranch exhausted, just in time to see Lonnie off in February for his hitch in the army. She fell suddenly and desperately ill. Only her toughness and the faithful tending by her friend, Irma Ledford, made her survival happen.

Lonnie graduated from college in 1958, and enlisted in the army. When he was discharged after his two years were up, he visited his brother Charley, Jr., and his family in North Dakota before heading for Hells Canyon to run the ranch.

In 1960, Tim graduated from college. Charley and Evelyn's dream and bitter struggle to get their children educated was realized. Tim was immediately drafted into the army.

On the dark side, Dinosaur National Monument expanded in 1960, completely engulfing the Mantle grazing permit and all its private land completely within its borders. Rules and regulations, harassment, and cuts in his permit left Charley bitter and feeling defeated, as discussed more fully in later chapters. In 1965 he signed the ranch away to his five children and went to chase his dream of seeing the ranch country of Brasil before he died. Evelyn encouraged him to go, even though she felt he would surely perish down there. She felt that it was his dream, and he had a right to do it.

He asked her to go with him, but she knew it would be a camping trip in the roughest kind of country, and she could not live through that kind of hardship any more. Evelyn gathered up a few belongings and went to live with Queeda in Boulder for the winter, fully intending to return to the ranch when she was able.

2

Memories, 1958

TIM MANTLE KICKED UP A TRAIL OF DUST behind him as he trotted off on his journey down to the headquarters of the Mantle Ranch. He was enjoying the fresh, crisp air of Blue Mountain and the feel of the good horse under him. Tim was only twenty years old, but he already had the rugged good looks of a Western rancher. He rode his horse like he was a part of it. It was the summer of 1958, and he had just gotten home from his second year of college at Colorado State University.

Soon the trail led into rocky, steep terrain as it headed down the mountain. Deep shadowy Hells Canyon yawned to his left. He was riding the age-old trail he had ridden from childhood. It wound around the west side of Martha's Peak, becoming quickly a narrow path. Tim let his horse pick its way through the jutting rocks and stumpy cedar trees that invaded the trail. As he rounded a bend and came into the first view of the land far below, he stopped to look and to remember.

The Mantle Ranch had been homesteaded in 1926. His mother, Evelyn Fuller, was a young girl from New York who had married a penniless cowboy, Charley Mantle, fourteen years her senior. They had joyfully, full of love and hope, ridden this trail. It had led to a life of hardship and nearly superhuman accomplishments for them.

Far below, Tim saw the beautiful golden cliffs of the Yampa River Canyon glowing in the sunshine. Deep down there in those canyons his family had grown up. There were no roads, so everything was done with horses. Even groceries had to be carried in on a packhorse these forty miles from the grocery store. There had never been and still wasn't electricity, telephone, or running water.

As each of their five children were born, Charley and Evelyn provided the best they could for them, and it was adequate in every way.

Food was grown, raised, hunted, fished, and always there was enough. Each child had been educated through college if they wished to go. Tim winced as he remembered the hardships his mother had endured and the heights she had risen to in the education of her children. His father had managed to always provide what was needed. He was the best cattleman Tim had ever known.

Charley had taught his four sons all the ways of the West. They could ride anything that walked—or nearly anyway, and raise cattle in the worst of conditions. The two youngest boys were educated with college degrees, and boy, did they know how to have fun! Tim smiled as he thought of all the rodeos he and his brother, Lonnie, had won big-time in lately. They were both on the college rodeo team, and also entered all the local rodeos on the Western Slope.

Tim barely remembered his oldest brother, Potch, who had left home after a tizzy-fit when he was sixteen, and Tim had only seen him a couple times since he left. He knew Potch had rodeoed professionally for many years, and the rumor was that he now was married and living in Montana. Tim wanted to go see him and get to know him. Pat, his next oldest brother, was Tim's hero. Pat was always so helpful and protective of the younger kids. Pat had been briefly married to Sue Woolley and had a son who lived in Meeker. Steve was his name, and Tim smiled as he thought about how he loved him almost like his own. Pat had just started his own rodeo string called the 7-11.

Evelyn and Charley Mantle – 1958

Tim Mantle at a rodeo

Tim thought of his sister, Queeda. She had married a Texan named Rex Walker she had met in college. They had just moved to Boulder, Colorado, so Rex could finish his master's degree in Geology at Colorado University. They had a baby girl, and there was some talk of Rex and Pat going into business together.

Tim's brother, Lonnie—who the family called "Nav"—and he had always been inseparable. They had "cowboyed" together all their lives. They had gone to school together through some pretty tough years at home, then both had graduated from Wasatch Academy, a boarding school in Utah. Now they were in college together at Colorado State University in Fort Collins. Lonnie would graduate the following spring and go into the army. Tim would not graduate until 1960, when he, too, would be drafted.

The two boys had made a deal between them. Dad was getting old and crippled up, so he couldn't do much ranch work anymore. Evelyn had been sick and was weak and frail, so she needed to get off the ranch and heal. The deal was that Tim would run the ranch while Lonnie was in the army. Then Lonnie would come back and run the ranch while Tim was doing his hitch in the army. After that they would likely run the ranch together. Tim smiled as he thought about how great it would be to have his brother back with him. He envisioned them letting their parents rest and enjoy the ranch after so many years of hard work and sacrifice.

Tim stopped his reverie and nudged his horse back to his careful traveling down the rocky trail. He had to get home and check on his mother. She wasn't well since getting back from Texas, where she had gone to be with Queeda when the baby was born, then helped in the frantic packing and trip to Colorado with that tiny baby. Tim also had to be sure the lawn and Mom's garden were watered, and that the creek water irrigation dams were reset to different locations in the orchard.

3

Charley, Jr.

NOBODY KNEW WHERE POTCH—CHARLEY, JR.— the oldest son of Charley and Evelyn, was. He had left home when he was sixteen. Reports came in occasionally that someone had seen him riding rough stock in rodeos. Evelyn, always afraid he was in some trouble or sick, or even dead, was always greatly relieved by such news. One year he made the top fifteen in saddle bronc riding in the Pro Rodeo Cowboys Association. He got the huge honor of riding in the national finals in Madison Square Garden.

The wild young man had been riding in a rodeo in Oregon. He had become good friends with another cowboy who was riding rough stock the same places he was. He accepted an invitation from his friend to go to his family's home for supper. The friend had a lovely sister who immediately caught Potch's eye. He decided to take in all the rodeos in the immediate area rather than taking off for some more distant places. The girl's name was Marmion Hart, and her family had been horse ranchers in Montana.

Sadie and Walter Hart raised horses by the hundreds. Unfenced, untrained, and wild as antelope, the horses had to be rounded up each year to sort and ship. The ranch area was between the Little Missouri, coming out of North Dakota not far from the Theodore Roosevelt National Park, and the Yellowstone River to the west. Walter had old cowboy friends living along these river areas that he traded horses with, and he kept building up his herd. Eventually he ran five different stud bunches—a bay, a pinto, a dark palomino, a light palomino, and a pure white part-Arabian.

This was the beginning of the palomino craze. Pintos were popular, too, and Walter had some of the best. He saved his filly colts.

He used mostly sorrels and bays. He was able to produce a good percentage of much-coveted pure white horses from his white stud.

This was strictly the prairie. He ran a lot on some national grazing reserve land that was located next to his property. It kept everyone busy, keeping his stud bunches separated for breeding purposes, and also to keep them from eating each other up. This was during the late thirties and early forties.

There were a lot of old cowboys living in that part of the country, and Jack Eaton was one of them. Jack's headquarters was his ranch outside of Glendive, Montana. He was a throwback to the old tough cowboys of the West that history is made of. He would later become very important in the life of Marmion and Charley, Jr.

Walter was very good with livestock, helped all the neighbors with theirs, and was the resident veterinarian for the neighborhood. When they needed help, someone would say, "Go get Walter, he'll help us. He'll know what to do."

Most everyone in the area had migrated from Minnesota. The older brothers in the Hart family came to the country first, and they bought up old farms and homesteads. Everyone was basically a farmer.

Marmion's mother, Sadie, came from Minnesota as a schoolteacher and taught the country school—that's how she met Walter Hart. She taught the older kids in the families of his brothers and sisters and other neighbors' children and boarded with Walter's older sister and family. Sadie was a joyful, fun person and was always popular with the young people of the family.

As for all of Walter's horses, they made a living for his family. He broke them and sold and traded them for things he needed, broke other people's horses, and broke teams, as folks there still farmed a lot with horses. There was another family that lived close by, who had young men in the family. They all worked together and got the work done. They rode together and would put on rodeos for the community for fun.

Sadie Hart's family made the gut-wrenching decision to move to Oregon for health reasons. Walter's health was not good enough to continue the rugged ranch life and harsh winters that Montana demanded of him. They shipped the horses by railroad from Webo, Montana, to The Dales, Oregon, in the year of 1946. The family all moved there. Walter sold most of his horses after arriving there, as it was not good horse country like eastern Montana.

A Montana ranch girl at heart, Marmion missed the ranch life she knew. Charley, Jr., finally mumbled to Marmion, "Hey, wanna get married?" The answer was yes, and they were married in Hood River, Oregon, in 1955. They moved almost immediately to Glendive, Montana, where Charley got jobs on the oil rigs for a short time, but he was happiest when he could make a living the cowboy way. He became well known for the livestock man he was, and he worked for many of the ranchers in the Glendive area. Jack Eaton heard of him and came to see him with a deal. Jack had bought up many small ranches in the area and ran his outfit under the HOT brand, and called it the Hot Bar Ranch. Charley, Jr., jumped at the chance, and it proved to be a good job and paid good money.

Marmion and Charley, Jr., lived in the Glendive area until August of 1957, then they moved to Dickinson, North Dakota. Charley knew an Indian man named Joe Chase, and they wanted to run cattle together around the reservation there.

On December 20, 1957, their first child, Cammie Lynn Mantle, was born in Dickinson, North Dakota. What a wonderful miracle in their life she was. Gambling was legal in North Dakota, and Charley, Jr., could make a living as a card dealer during the winter months. He continued to work for Jack in the summer and fall.

Jack's ranch was in the jumbled, rugged badlands and had many, many wild horses on it. They kept the feed on the range picked clean, and Jack wanted the wild horses rounded up so he could have that range for his cattle. The gathering of these wild horses fell to Charley, Jr. This project began in the fall of 1959 and continued for the next three years.

It wasn't over yet, but we must go back to the Mantle Ranch for a look at the rest of the family.

4

Lonnie (Nav), Queeda, & Pat Mantle, 1953 to 1959

AFTER HE GRADUATED FROM HIGH SCHOOL, Lonnie attended Colorado State College in Fort Collins. While he was there, the school's name changed to Colorado State University. He majored in Animal Sciences, which related well to what he wanted to do for the rest of his life, which was ranching. He pretty much paid his own way through school. His dad paid his tuition, but food and lodging were his problem. He and three or four other friends rented a house together. Jack Haslem, his old friend from Blue Mountain, was already in school there, and through him, Lonnie got a job cleaning stalls and such around the college ag department.

Lonnie made the college rodeo team during his freshman year and was a top performer for the Aggies all through his college years. Jim Karman, one of his roommates, was responsible for Lonnie moving to Riverton, Wyoming, later on. Jim had a job in Riverton, and Lonnie moved there to be near him, and ended up spending the rest of his life in Wyoming. Jim, the other two roommates, and Lonnie have remained fast friends through the years. During their college years none of them had any money, so they shared expenses as best they could. One quarter, Jim furnished a 100-pound sack of pinto beans and Lonnie furnished the venison. The meat arrived at their apartment as whole deer! They cut it up on the kitchen table and the other roommates scraped up enough money to store it in a meat locker.

Lonnie got a scholastic scholarship along with his rodeo scholarship, and that helped a whole lot with expenses. He graduated from college in June 1957, and enlisted in the army, as he was going to be drafted anyway, and reported for duty in February 1958 for his two years of military service. He started up a hair-cutting business in his barracks and did pretty good with it—until, that is, his first black customer came in for a hair cut. Lonnie plunged his clippers in and started work like he always did, and to his horror discovered that his clippers were so entangled in his buddy's curly hair that he couldn't get them out! After a lot of pulling, clipping, balding, yelling and cussing from both of them, he finally got the clippers out. Pretty sure he was going to get a butt kicking, Lonnie was relieved when the guy left mumbling something about he just wanted to get out of there.

During his tour of duty, Lonnie was sent to Germany. He made the most of that, and he and a buddy bought an old car and did a lot of touring around. He had always been interested in photography, so he took the opportunity to buy a really fine Leika 35mm camera. That camera was his pride and joy. He studied and practiced photography all his life, and he became an expert photographer.

When he was discharged, Lonnie went by Dickinson, North Dakota, to visit his brother Potch. After a nice visit, he headed for Hells Canyon to take his turn to run the Mantle Ranch. Pat had been taking care of things, but he needed to be free to run his 7-11 Rodeo Company he had started. Besides his rodeo string, Pat Mantle and his brother-in-law, Rex Walker, had formed a company they called "Sombrero Ranches." They planned to rent out riding horses to summer camps and church groups and planned to build a stable to operate themselves.

※ ※ ※

Tim had been in college in Fort Collins and graduated in June 1960. He was drafted almost immediately. Faithfully, the boys ran the ranch for each other over those busy years and gave each other the chance to serve their time for their country, as well as tend to their individual businesses without the ranch being abandoned. Evelyn hung on at the ranch to cook and make a home and garden for them.

※ ※ ※

Pat and Rex rented out their first dude horses the spring of 1959. They both took some carpenter tools and a few dude horses and moved to Estes Park. They leased trail riding privileges from Miss Muriel

McGregor, the lady who owned a big ranch on the east side of Estes Park. They built a small stable to rent horses out of. They lived in it while they built it. It was open for business on July 3, 1959, and in the company of swarms of flies and happy customers, Sombrero Ranches had its beginnings. They owned fifteen horses, which they rented out for $1.50 an hour. They made $225 that first day. The young entrepreneurs were elated. One dollar and fifty cents per hour was a lot of money, and they still got to keep the horse to rent out again.

That spring Rex and Queeda had bought the small acreage next to their rented house at 1300 Cherryvale Road in Boulder, Colorado. The old farm house on the place was a wreck, and Queeda was left to move into it by herself. Rex and Pat were too busy with the launching of the new business to help.

Queeda, seven months pregnant, wanted to be in Estes, but knew she didn't have much time to get moved in. The daughter of a neighbor, Ray Oram, came and helped with toddler Cindy, and helped get them moved inside the walls of the new house. She was wonderful and was deeply appreciated for all her help. In August, faithful Evelyn appeared to help her daughter with the birth and care of the baby. Justin Ross Walker was born on August 9, 1959. Evelyn had to hurry on home, as Charley wasn't well, but she promised to come back for the winter. Evelyn wasn't well, either, but her own health seemed to always be her last consideration.

5

Charley Goes For It, 1960

JOY AND ENTHUSIASM SURGED THROUGH CHARLEY! It seemed that forty years had fallen from his broken, aching sixty-seven-year-old body. He could finally leave this damned place behind that he had loved but been tied down to for thirty-four years. Dinosaur National Monument had just expanded their borders once again. Now, all the Mantle Ranch and its grazing permit land were located within Monument borders. Charley had just told the boys he was leaving the ranch to them, and that he was leaving the ranch forever. Let his grown kids battle the sneaky government bastards that wanted to put the ranch out of business! His boys were the best cowmen he knew, and two of them even had college degrees, if that meant anything. If the Old Woman wanted to keep on batting her head against the wall, let her do it. He was going to Brasil!

He dreamed of riding through rolling plains of grass up to his horse's belly as far as he could see, with great herds of fat cattle grazing it. He could hear the black-eyed, sinewy cowhands yelling and cussing in the universal language of cowboys as they rode mounted on their shining horses, doing the things cowboys always do when set free. He was sure his endless studying had made him able to talk with them, and he sure as hell could ride with them.

He packed a duffle bag with an awl and leather strings to repair leather with. He put in a change of clothes and an extra pair of socks. He rolled up a wool blanket as small as he could make it—his saddle would do for a pillow. He had his pocket knife in his pocket. He bound up his saddle and two saddle blankets, a hackamore, bridle, and his lariat and lashed them together so they would fit in a gunny sack. There was plenty of room left in his duffle for traveler's checks and the papers he would need. He stuck in a stubby pencil, his reading glasses, a bottle of aspirin, and a bunch of chewing tobacco.

He gave Evelyn a bear hug and climbed in the yellow Jeep. He jumped it out of the yard like he always did; he revved the old Jeep up real good, then yanked his foot off the clutch as he slammed the other foot down on the gas pedal. Evelyn had just wished him well and said to have a safe trip. She felt that a man who had worked so hard all his life had a right to try to fulfill his dreams.

Charley went to Vernal, Utah, to get a passport. He found it wasn't possible. He would have to find somebody who could swear he was born because they had been there at the birth. It didn't seem likely there could be anybody that old still around. Finally, it was agreed that he could use his army service papers. They were in a safety deposit box at the bank. He got a copy of those papers, checked out most of the money in his bank account, bought traveler's checks, and caught the Greyhound bus for Denver. There he called Queeda to come get him, and she drove him to Boulder.

He finally was able to get a passport. He bought a plane ticket to La Paz, Bolivia. He was disappointed that his boys couldn't go with him, but didn't dwell on it. There was no hesitation or foreboding in him, just excitement. He was on his way!

Charley had gotten together a whole lot of information on Brasil. He knew where the best cattle country was and who had settled the country, and he knew the routes he would take to get there. He understood that the language of Brasil was Portuguese, but he figured it would be near enough to Spanish that he could be understood enough to get around. He particularly wanted to investigate the King Ranch of Brasil , because he had a lot of respect for them as cattlemen, based on what they had accomplished in Texas. He also considered their herd as the probable seed for the herd he would run, because they were bred to withstand the heat and the insects that infested hot, humid tropical country.

Charley flew out of Denver on July 2, 1960, headed for La Paz, Bolivia. Everything was new to him in the world of civilization. The airport was teeming with people all in a hurry. They checked and rechecked his passport and ticket. At the desk, a ticket agent could see that the old cowboy was having trouble coping, so he called a courtesy cart to pick Charley up at the desk and take him straight to his departure gate. On the plane the stewardesses fussed over him. When one of them took his new hat to stow in a compartment above his seat, he coached her about placing it resting on its crown, and placed where it wouldn't be smashed by anything. The man in the seat next to him invited Charley to trade seats with him so he could sit next to the window. It was obvious that this was the first airplane trip for this cowboy, and he was very excited about it. The stewardesses plied him with refreshments.

The big plane roared into the sky, and Charley had on his most stoic face to hide his apprehension. It was a bright clear day, and Charley stared fascinated at the mountains and plains so far below. He commented to the man seated beside him that this sure got you there a lot faster than a horse. Deep in conversation, the time flew by, and when the plane landed in Mexico City, the man reluctantly got off the plane, leaving this interesting old cowboy to finish his search for a dream.

Back in the air, Charley looked out the window and saw hundreds of miles of jungle with an occasional river winding through. Finally, he saw snowcapped mountains and knew he must be nearing Bolivia. It seemed to him that the plane sailed rather low over the mountains, then it surged to a higher speed as the nose tipped down, and the mountains flew by level with his window in a blur. Suddenly he could feel the plane shudder as if huge brakes had been applied. He was driven hard against his seat back. Miraculously, it seemed, the plane landed safely on the runway and taxied to the airport building.

La Paz is the world's highest capital and is described as "the city that touches the sky." Surrounded by even higher snowcapped peaks, the airport sits in a bowl at 13,450 feet. Planes have to fly at very high rates of speed in the thin air. Charley knew all this from his studying, but it still was a hair-raising experience for a cowboy from Hells Canyon.

Charley got some American money changed into bolivianos, the Bolivian money. He also bought a few Brasilian reals. He saw a bus parked outside that said La Paz on the side of it. He got on the bus, hoping it would take him to the city, which lay at 3,600 feet far below the airport. The bus took a crooked, winding trail down the mountain and into the city.

Bolivia is known as the "most Indian" country in South America. Charley had grown up around Indians, and was able to talk using the universal sign language. The people were friendly and helpful to him. He got a room in a cheap hotel, ate at a little restaurant, and, as he drifted off to sleep, gave a deep sigh of victory and satisfaction at the outcome of his first day.

At dawn the next day, Charley was up and excited about seeing La Paz. After breakfast he found a tour bus driver who spoke good

English. He would have time for a tour, and then the bus would be ending up at the bus station in time for him to catch the next bus out of town. He sat right behind the driver so he could talk to him as he enjoyed seeing the ancient old city. He had his picture taken with an Indian woman in the characteristic flat black derby. He sent the picture home in a letter later on. The thin air was beginning to exhaust him by the time he got back to the bus station. He also had been cold ever since he got here. He was pretty sure this was the last "good" air he would breathe for the rest of his trip, because he knew he would be in a steamy land. He bought a stamped postcard and wrote a short note home. He knew the family would be worried about him.

On July 4, 1960, Charley sent his first postcard home.

La Paz, Bolivia
Dear Queeda,
I am leaving La Paz tonight at 10:30. Will write you from Corumba.
It is cold here. Got on all the clothes I got. Dad

Charley was exhausted and wished he had waited until the next day to take the bus. Well, maybe he could get some sleep on the bus. He got on board and settled in. It would be dark for awhile, then he could get a look at Bolivia. He was on his way to Corumba, Brisil, and he knew it was a long trip just to get out of Bolivia.

The bus took Charley down some mountain roads that were unbelievably steep, poorly maintained, and with one sharp curve after another. Some of the turns were so sharp the bus had to back up to make them. He learned later that this was called "The Road of Death." Finally they arrived into a jungle-choked flat plain. The clearings were filled with tall waving grass, just as he had dreamed of.

The first place the bus made a rest and eating stop in this jungle country was a shock to Charley. The bus wasn't air conditioned, but just its movement kicked up a little breeze. Here, as he stepped out of the bus, the air was still and he felt like he was in a steaming kettle. Hordes of insects large and small moved in immediately on the passengers and began to feast on them. Charley pulled down his hat and turned up his shirt collar and went in the little shack looking for the restroom. It was the filthiest little bathroom he had ever seen; he almost vomited. He turned around and stomped back outside and into the jungle a short way and relieved himself. Hungry, as it was already mid-afternoon, he went back inside to see what they had to eat. He settled for something stringy and dark brown slopped on a tortilla and a bottle of hot Coca-Cola.

After he ate, Charley went outside and looked around. An old man was sitting on the shady side of the shack, hollowing out some dried gourds. Charley signed to him, asking what he was making. The man signed back, explaining the gourds were to carry water in. They dickered for a price, and Charley paid him in reals, the Brasilian money he had bought in La Paz. The old man shook the big gourd clean, then quickly whittled a cork for it from a tree branch. He beckoned Charley to follow him and took him to a stone structure nearby. With a rope and bucket stored nearby, he pulled up a bucket of blessedly cool, clear water. Charley took a long drink from the bucket, then held the gourd while the old man filled it up.

Back on the bus, they wound their way through the countryside, and Charley watched eagerly. He had come here to see pastureland, not jungle, so he was disappointed in that way. Occasionally he would see a herd of cattle and a few horses. His studies had shown him that Corumba, Brasil, his destination, was a town in the state of Mato Grosso. Mato Grosso stretches to the north, and the whole state's economy is based mainly on farm products, mostly cattle-raising. Corumba is one of the main entrances to the huge Pantanal wetlands region. It is an important inland shipping port, since it is situated right on the giant Paraguay River. Now he knew why it was called "The Southern Thick Forest."

He realized that he had miscalculated the distance from La Paz to Corumba because the state of Mato Grosso was so big. He was just too far south, but although it was a much longer ride than he had thought, it had been a fascinating trip getting here.

He got his passport stamped with a visa for Brasil when they crossed the Bolivia/Brasil border. Then he got a taxi ride into town from the bus station, and the taxi driver took him to a small, inexpensive hotel. His body ached from the long trip in the bus. He was hungry and very tired. He managed to get a meal at a little restaurant right by the hotel, then he crawled into bed.

6

Corumba

CHARLEY WROTE a letter home:

July 11, 1960 Corumba, Brasil
Well I finally made it to Corumba. It is a hell of a place. I don't think they eat here, just drink and gamble. This river is really big. I saw a ship leaving here tonight for Buenos Aires with a load of cattle. I saw some good grass country in north east Bolivia, but I don't like that country. Their money is no good—12000 to the dollar and mostly Indians and they are all dope heads.

  I don't know where I will go from here. I am going to stick around here for a few days to see if I can get some information. Will write again before I leave here.

  I am feeling fine. Only trouble I have is getting chewin' tobacco. They don't have any and I am chewin' natural leaf now that I bought from a Indian.

  Saw a few horses around here on the streets. They are the damndest lookin things I ever saw. About the size of Cracker Jack [a Welsh pony], *and their saddles I don't think they can tell the front from the back.*

  Well I am going out to see if I can talk enough Portuguese to get some supper. Dad

Charley only spent a day in Corumba. He didn't like it and felt like it wasn't a safe place to be. It was a big, dirty, heartless industrial city, and it was even worse for him because he couldn't speak the language. He was able to sign enough to get food to eat and to buy a bus ticket to Campo Grande, which was about 200 miles to the southeast of Corumba. He was soon to learn just how vast this country was.

Most ranchers owned airplanes to get around in, as their ranches and numbers of cattle were so unimaginably big.

In a letter written July 27, Charley told of this trip:

> *I went from Corumba to Campo Grandy, spent about 4 days. I met some fellows there that have a hundred and twenty thousand cattle. Old man and four boys. One of the boys took me in an air plane and showed me some country. We flew to one of the ranches, then took a jeep. Spent five days there. He said we would visit some of the neighbors.*
>
> *It took us all day and part of the night to get to the first neighbor. I didn't know there was so many cattle in the world. I think I saw 300,000 cattle. The country was full of wild hogs, ostriches, turkeys, deer, alligators, monkeys, and I didn't know there was so many pretty birds, but nobody could live in that country but a native. It a low swampy deal only 6 to 9 feet above sea level.*

Charley took advantage of the chance to rest up at these people's nice ranch. They had running water and plumbing and electricity, all run by generators and solar power. He also enjoyed the good food, especially the beef. He hadn't had any good beef since he left Colorado. After seeing so many cattle, he was baffled at how the food could be so bad in restaurants. His hosts hated to see him go. They had enjoyed this handsome old cowman with the determination of a bulldog to find a place where his sons could ranch in peace and prosperity. He told them yarns of his cowboy days, and they told him theirs.

It was a grueling 700-mile bus trip from Campo Grande to Cuiaba. Charley told about it:

> *July 27, 1960*
> *Brasil (Cuiaba, Mato Grosso)*
> *I got here last night. I am trying to look it over as I go.*
>
> *I came back to Campo Grandy and took a bus to here. WHAT A RIDE! It is just a new road about 700 miles. We was two days and this country aint worth a damn. Lots of grass but all brush like south Texas.*
>
> *But about 75 miles back before we got here we got up in some high country out of the timber that was the best country I have ever seen. Just as far as you could see all ways from the road nothing but grass, and I didn't see one cow. I don't know what is the*

answer, maybe no water. I haven't seen anybody to talk to yet, but they say there are some Americans here so when I rest up a little I will see if I can find somebody. I think I am just beginning to get in the good country.

This place here don't amount to a damn. It is the capitol of the state, but I don't know why. I will try to find out. I don't like it. It is old and hot and dry and I don't know what their source of income is.

I don't know where I will go from here. I think I will have to take a pack outfit to see the country. Don't think there are any more roads. I think it is Indian country from here on. If you write me here I will try to make some arrangement to get it. I think I will be in this part of the state for some time. Dad

Cuiaba was founded with the discovery of gold at the beginning of the eighteenth century. It is the capital of the state of Mato Grosso. Cattle raising is its largest industry.

7

The State of Goias

THE LAST TIME WE HEARD FROM CHARLEY was forty anxious days ago from Cuiaba in the east central part of Brasil, in the state of Mato Grosso. He left there by bus and traveled directly west into the state of Goias. Goias lies wholly within the Brasilian Highlands, which are located in the center of Brasil. It is a huge, almost level plateau that runs east and west. It stands between 2,500 to 3,000 feet, with the highest point being 5,500 feet high.

To the south, Goias is drained by a Parana River tributary. To the east, it is drained by the Sao Francisco River, and northward the state is drained by the Aranguaia and the Tocantins Rivers.

The state is covered with a woodland savanna, known in Brasil as *campo cerrado*, with tropical forests along the rivers. The climate of the plateau is subtropical. Temperatures range only between 71 and 80 degrees Fahrenheit. It rains a lot during rainy season. Goias is a very large cattle-producing state. Its capital city, Goiania, and another town, Anapolis, were and still are centers for meat-packing and food-processing industries.

Charley wrote home:

Barrado Garca
Sept.__ (I don't know what, I can't find a calendar) [Sept. 5, 1960, on the envelope]
I am in a little town called Barrado Garca. It is on the border between the states of Mato Grosso and Goias. I think I am the only American that was ever here. I am a funny lookin thing to them. I can't understand them. They think I am deef. They scream at me.

I don't know whether my letters are getting out or not. I never have got a letter. I met a fellow from Oklahoma that was going home. I asked him to write to you or Pat and tell you I was O.K. Don't know if he wrote or not. I haven't been in many places where I could write, but I think the roughest is over. I am heading east from here. I think it is a little more civilized, and I can write more. I won't go into any details because I don't know if this letter will get out of here or not. No use of you writing till you hear from me again because I don't know where I am going.

I will leave here in the morning. I am tired, but otherwise I am feeling good. Dad

Charley got a hotel room and holed up to rest for a few days. He got an old lady to wash his clothes, and he stitched up a few tears and rips in his Levis. His saddle was getting stiff from being wet so much, so he greased it. He found a few people who spoke a little English, but there was nobody to satisfy his ravenous craving for information on cattle ranching in Brasil .
Charley's next letter:

September 15, 1960
Anapolis, Brasil
Well I finally got to Anapolis. I don't know if you have looked at a map enough to understand what I am talking about or not, but there is a divide that runs all the way across Brasil east and west. It is the watershed between the Amazon and the Paraguay and the Parana rivers. They are biggest and main rivers of Brasil. I tried to follow that divide all the way to here. One place I couldn't make it (but I am going back).

I really had hell. There are no roads; I have rode everything; these little South American ponies (Cracker Jack would be a big stud down here), little mules, old worn out trucks, jeeps, boats made out of logs. I have eat wild bore meat with Indians, slept with niggers, and eat flies by the bushel, but I enjoyed it all. I saw more grass and more beautiful water than I thought was in the world.

I am going to stick around here and Brasilia for a while and get same legal information. This town is full of sharks. Everybody you meet wants to sell you some land. I met a bastard from New York this morning. He is in the real estate business; He don't know any more about the country than Cindy (3 yrs. old) does and about 95 out of every hundred are just like him, but I think they are hookin a few saps.

I had a real pleasant experience the other day. I stopped at a little town about 300 miles west of here called <u>Ria Verda</u> and met a Presbyterian missionary by the name of Charles Sterout Billing. They took me home and I never was treated so nice in all my life. I had been about 30 days on rice and flies and hadn't seen anybody I could talk to. She fed me mashed potatoes and gravy, fried meat, eggs, everything I hadn't had since I left home. They took some pictures and said they would send you some. If they do I wish you would write them a nice letter and tell them how much I appreciated it and send them some pictures of the kids.

There is a tribe of Texans about 200 miles north of here. I think I will get around and talk to them sometime. They have been here about two years and should be onto ropes pretty good. That is, about the legal stuff.

I think if you write me Anapolis, Goias, Brasil I will get it. Write soon as you can. Write me ever few days. Dad

One month later he finally got a letter from home. It refreshed him a little bit, as he was tired and lonesome for family. Being hard of hearing made it terribly hard for him to communicate with anybody. The people didn't want to be patient with him, so he lived in a world alone.

Charley wrote:

October 10, 1960
Anapolis, Brasil
I just got your letter about ten minutes ago. It shore made me feel good. I was kind of down in the dumps. Haven't seen anybody I could talk to for about two weeks. There is an American here, but he lives about fifty miles out.

I just got back from the damndest experience I ever had. I wanted to go north from here. I got on a old rattle trap bus. I didn't know where it was going, but it was headed north. We got out about two hundred miles and it fell to pieces. There wasn't no ranches, towns, or anything. There was a nigger on the bus, so he took me by the hand and led me about four miles out in the jungles where there was a bunch of niggers living in a little Africa shanty. They had lots of parrots, monkeys, kids. All they had to eat was rice, but they made me welcome; the first night I took my hammock out in the brush and done all right, but the next day it set in to rain, and man did it rain, so I had to move in the shanty, niggers, monkeys, parrots, and dogs. We all eat out of the same rice pot, then the next day one of them died, so we had a funeral. We made a coffin of sticks and palm leaves, and dug a grave with a hoe. I was stuck there six days, then an old man and a little boy come along with an ox cart, and I got a ride with them two days. Then a truck came along and I hooked a ride on it and made it back here. I am going to try it again on a different route.

I am going north toward the Amazon. They say it is awful hot down there, but I want to see it. On this divide where I have been most of the time is a real nice climate, not near as hot as it gets in the canyon at home. I don't know what I think of the country yet. It is just so damn big and empty it is hard to realize. One thing I know, you can raise cattle by the millions, but it will take ten or twelve outfits to swing the deal. I will know more when I get through lookin it over. All the Americans there is in this country are away south of here; the King outfit, and all the Texans are down there. One American here, but he is not a cow-man, just old Texas cotton farmer.

I met a fellow the other day. He is a Spaniard, said he has been in this country three years. He has a ranch about three hundred miles north of here. He says it is a good cow country, and would show me around, so I think I will look him up. He is a pretty smart hombre; I think he knows what he is talking about. The most of these natives don't know straight up, not near as smart as Mexicans.

Just keep writing me here. I think I have got the people at the post office so they know me pretty good, so maybe I will get some of the letters.

I would love to see the kids. Dad

This letter upset his family at home very much. A sense of foreboding fell over everybody as they thought of him in the deep, damp jungles all alone. Visions of crocodiles, giant snakes, and cannibals were in the dreams of all the family. Nobody at this point thought they would ever see him again. As a matter of fact, this part of his trip did lead him into some tough times. You would have to know ranching and love it to know the excitement that drove Charley on, caused by all these cattle and beautiful rangelands.

8

Cowboy in the Jungle

CHARLEY STRUCK OUT NORTH ON A BUS, carrying his saddle and his valise. He was on his way to the Spaniard's ranch. About 300 miles north of Anapolis there is a town called "Paragatu," on the very border of the states of Goias and Tocantins. It is where all the people in the interior beyond shop for supplies. The river Tocantins flows to the north into the Eastern Amazon Basin, and on out to the Atlantic on the northern side of Brasil.

Cattle was the main industry of the area. The closer Charley came to Paragatu, the less of the vegetation of the *cerrado* existed, and the more it became tall waving grasslands. The temperature was rising, and Charley's sinuses started giving him trouble. He had a constant headache, but it didn't stop him for a minute. He was really disappointed to find that the Texans' ranches were way south. He wasn't going to have time to visit them. The rainy season was about to begin, and he knew he would have to get out of Brasil .

He followed the directions given him by the Spaniard and found his ranch. It was a beautiful grassland, but tough to raise cattle on. He had Zebu cattle, a breed well adapted to the tropics, but even they were bit up by insects and grubs. Besides, it was not run like the ranch he was looking for. The cattle were put in the corral every night and let out in the daytime. He enjoyed his stay there, and thanked the family heartily when he left.

He got on a bus going north, and all the things you read about in his next letter actually happened to him. Only much later did "The Rest of the Story" come out. The master yarn spinner couldn't keep it secret—it was just too good not to tell.

He traveled through beautiful country filled with great herds of cattle. He got off the bus at a ranch. Nobody spoke English, but he bought a horse from them and asked if he could work with them for a few days. He had a great time, and they were greatly impressed with his artistry with his lariat rope. When moving cattle across a river one day, he got to see a whole new danger he had never experienced. The men came to the river and left the cattle bunched and held there by some of the cowboys. The rest cut out one old cow and dragged her up the river about 100 yards. They turned her loose, and the wild old cow charged into the river. Then they shot her. Her blood mingled with the river water, then suddenly the water looked like it was boiling. One man rode up beside him and said, "piranhas." Charley watched in horror until only the skeleton of the cow remained. He finally noticed that his horse was in a frenzy to go, and he realized the men had left him. He saw them crossing the herd of cattle down the river while the evil fish were eating upstream. He quickly jumped his horse into the river and crossed with them. He knew for sure he wanted to stay out of any rivers for the rest of his trip.

Deciding it was time to move on, Charley found one of the men who knew the country, and they drew a map together. All the rivers and trails were in it, and it led to the next big ranch. The route laid out on the map would take him through the country without having to cross any rivers. That was important to him after seeing the piranhas eat a whole cow. He wished he had his pistol, but he didn't, so he supposed he would have to rope and fish for his food. It didn't look far to the ranch.

He declined an offer from the ranch foreman to send somebody with him and took off. Soon the jungle became deeper and thicker. It was eerie and dark. The trail was well-worn from years of cattle trailing through there, so was easy to follow. However, the intense heat and 100 percent humidity made his joints ache, and worse yet, his headache became a pounding torture. He could hardly see. The horse was good about following the trail, like he had been on it before, so Charley tried to hold down his mounting panic and just ride. He felt like he was being eaten alive by the insects. Darkness began to set in. Bats began darting around his head, chasing mosquitoes. With a great deal of relief he smelled smoke, and soon he saw the flames of a campfire. "Must be at the ranch," he thought.

He pulled up his horse and looked in disbelief at what he saw before him. A group of nearly naked dark-skinned Indians were sitting around a campfire. They barely acknowledged him, so knowing nothing else to do, he dismounted, tied up his horse, and approached them. They grunted and motioned for him to sit down. They had long matted hair and didn't smile. Through his swollen bloodshot eyes, hammering headache, and blurred vision, he saw that they were having their supper. He was fascinated watching them. They had long spears. They would swat bats to the ground as they fluttered over, then skewer them on the spear and suspend them over the flames until they were cooked to their pleasure. Then they would blow on them until cool enough to eat and pop them in their mouths. He was offered several bats on a stick, but declined.

One man got up and walked over to him. He put his face right up close to Charley's face and stared intently at him. Then he turned around and disappeared into the jungle. Several minutes later he returned with some green leaves. He dried them over the flame and pulverized them in his hand. He motioned for Charley to take them in his own hand and snuff them up his nose. He did, and it set off the worst coughing, sneezing fit he had ever had. For what seemed like forever this went on, and the liquids drained out of Charley's nose in streams. Finally it all stopped abruptly and his headache was gone and his swollen face had gone back to normal. He didn't know what it was, but he knew he had to have some, so he asked the man, and the next morning there was a pile of the dried leaves set aside for him. He wrapped them carefully in a cloth and stored them in his valise. There was also a pile of bright red tiny peppers that he knew were terribly hot, but he had admired them as the Indians ate them with their bats, so he took them with him, too.

The Indians had left, and Charley found his horse unsaddled and eating the last of a big armful of grass. He mounted up and went on his way. Charley felt so good with the pressure relieved from his sinuses that he once again was able to enjoy just being able to ride in this beautiful green jungle. About noon he began to have pains in his stomach, which he recognized as a gall bladder attack. He had had troubles before with this and had hidden it as best he could from Marmion and Queeda so they wouldn't try to haul him in to a hospital again. That was where people died, and he would rather die right here than in a hospital.

He began to see a few straying cattle around in clearings in the jungle, so he knew he must be getting near the ranch. The pain became a growing fire in his body, and the pain became so intense that he finally had to slide off his horse onto the ground. He lay there for a day and a half, wishing for death to ease that terrible pain. He drifted in and out of consciousness. His horse wandered on down the trail and luckily had made it to the ranch Charley was headed for.

The rancher's eighteen-year-old son, Jamie Roman, came backtracking the horse down the trail and found Charley near death, leaning against a tree. He got Charley on his horse, which he had brought back with him, and slowly led him to the ranch. There the good people got him to a doctor and nursed him back to health. They were in a rage at Charley's family for letting this old man come down into the jungles alone to certain death. Mrs. Roman wrote to Charley's family, expressing her anger and telling them his condition. The Roman family tried to get him to stay with them, but he insisted on moving on. His health was not very good yet, and he was weak from being ill. His gall bladder was not giving him great pain, but it still hurt.

Charley wrote home after this experience:

November 4, 1960
Little town north of Anapolis Brasil
Just got your letter. I just about cried about that little boy [Justin, Queeda's son had inhaled a crayon into his lungs.] That was really rough, but it could of been worse. I got back about a week ago from my trip up north. That was a bastard. Everything was moving with ticks; so hot your clothes would stick to you like a plaster; just about everybody has the fever. They said there was a leper camp just the other side aways and some of the lepers families lived in this village, and they come home to visit often. Of all the wilds of Brasil I've been in, that is the only time I got scared. I shore got the hell out of there.

The rain has set in. It rains every day. Everything is green and pretty, and every tree and bush has flowers on them. I met some people the other day that used to live in Greeley. Their name is Roman. Real nice people, but the old lady has flipped her dipper. They have been down here since 1948.

I hope Tim's army deal pans out like it sounds. He will be home by spring.

I am going to leave here tomorrow and go to a place called Formasa. It is north and west of Brasilia. They say there is lots of cattle up there. Think I will stop a day or two in Brasilia. They say it is a beautiful city.

I don't think there will be any use of you writing until you hear from me again. These bastards won't forward any mail. I will write again soon.

9

Charley Gets It Done

RELUCTANTLY THE ROMAN FAMILY drove Charley into town and put him onto a bus for Formosa. Unable to resist seeing one more real cattle country, he was eager to see if all he had heard and read was true—that Formosa was the center of a great cattle industry. Formosa, in the state of Goias, fifty miles east of Brasilia, is one of the most important cities in the area. It is a large producer of cattle and grains and is known for its waterfalls and natural beauty. It was a good place for Charley to rest up. He took short trips into the country and, sure enough, there were cattle grazing as far as he could see on the gently rolling flat plateau lands, covered with the *cerrado*-type feed. It was now entering the summer, or rainy, season, which is from October to April.

Charley began hearing familiar disturbing complaints in the street. "Cattle in the state of Goias are destroying the *cerrado* vegetation. Trees are being cut down to make pastureland for cattle, which is causing extensive erosion. Cattle are causing the erosion of the riverbanks." All of this was said to be causing extreme erosion and the danger of a water shortage in the state during the dry season, and there was a great clamor for the government to "do something" about the destructive cattle. Cattle seemed to be blamed for any malady that existed. Charley's blood turned cold at the sound of all this. He had struggled under government meddling in his livelihood all his life. The government had ultimately shrunk his ranch and placed impossible restrictions on the use of his land until he couldn't stand it any more. He had left the ranch in search of a better place to raise cattle. He didn't want anything to do with this country if it was going to be a battle against the government.

He packed his valise and his saddle and boarded a bus going east. Brasilia had just been made the new capital of Brasil this very year and he wanted to see it, but first he wanted to see the Brasilian Highlands, which run parallel to the Atlantic Ocean.

He crossed the mighty Rio Sao Francisco, and after traveling by bus for about 300 miles, arrived in a little town named Montes Claros. The main industries were beef and dairy cattle. As an item of interest, in 2007 Montes Claros had the largest factory of condensed milk in the world [owned by Nestle]. He settled in at a small hotel and as usual couldn't find anybody who spoke English. He went out to the airport and got it across to them that he wanted to take a plane ride to see the mountains and fly up the Sao Francisco River.

The pilot, soon seeing how interested he was in everything, began taking Charley for a look at waterfalls, beautiful mountain valleys, and high dry plains. They flew north up the mountain chain, then circled and flew back over the grazing land. It was a beautiful sight, and the pilot delighted in flying very low so Charley could get a good look. Charley could see that this land, as he had read, was just recovering from the dry season, which is always very brutal, with many fires and little water and feed.

Satisfied that he had seen the best Brasil had to offer a cattleman, Charley boarded a bus and rode it back west to Brasilia. He went to the airport to send a letter home and to buy a ticket. There was only one remaining thing that he wanted to see in Brasil , and that was the great Amazon River. His family was continually impressed at how he could get around in a huge city, and even a whole huge country where he couldn't speak the language, and accomplish anything he wanted to do.

Charley wrote home:

November 15, 1960
Brasilia, Brasil The following is a letter Charley wrote that was taken to New York by Pan American World Airways. It was mailed from the New York International Airport in Jamaica, New York, on November 17, 1960 at 4:00 p.m. in an "Unaccompanied Baggage" envelope with a 4-cent stamp on it.
Dear Dotter,
Well I am in the great city of Brasilia. It is some town – nothing much yet but it looks like it will be. I have been about three hundred miles East of here. I think I have got it pretty well looked over.

I have bought a plane ticket to Belem. From there I am going up the Amazon river to Manaus. It rains all the time.

I guess today Tim goes in the army. I hope he doesn't have to stay too long.

I have met a fine gentleman here. His name is reverend Douglas Charles. He wants to borrow some money. He has got lots of good deals if he can raise the money.

I am going to put this letter on the plane that goes to the states today so it will be mailed in the states. My preacher friend is going to give me some stamps.

I will write you from Belem. Dad

He handed the letter to a clerk at the Pan American desk, then boarded the plane for Belem. Belem sits on the northern coast of Brasil at the mouth of the Amazon River, where it pours into the Atlantic Ocean. He would be right on the equator, and Charley wondered what that would be like.

Charley enjoyed his ride high over the jungle in the beautiful, luxurious airplane. He got a window seat and gazed in wonder at all the many rivers he flew over. He knew they were all tributaries of the mighty Amazon. He thought of his beginnings in the little landlocked town of Vernal, Utah. He was orphaned at thirteen and had worked hard all his life, beating a living out of a dry land with a herd of cattle. Never had he ever dreamed he would get to see the Amazon River or South America.

He was satisfied!

Belem, which means Bethlehem in Portuguese, lies in the very northern part of Brasil in the state of Para. As the plane banked, Charley could see the coast and the Atlantic Ocean, then they were above a very large, wide bay. As they circled around Belem, he saw many small islands around it. The city lies on the estuaries of two rivers, the Tocantins and Para. Belen began in 1616 as a river port, with the mouth of the Amazon river a short distance to the north. He could see the rivers as they ambled through thick jungle and swampy land. Belem is known as "the city of the Mango trees" because there are so many of them growing there.

Charley landed in Belem, took a taxi from the airport into town. He had it drop him off at the pier where Amazon boat travel seemed likely to take place. He found a place that looked like it sold boat rides, and inquired inside. The man at the desk found a man who spoke English to wait on Charley, and he was relieved to be able to explain his wishes and also understand what was said to him. He felt "hungry" for spoken English. The man told him of several different trips he could take up the Amazon. He produced a map that showed the route to the destination of Manaus, which is where Charley wanted to go. Charley told him he wanted to get to Panama City from Manaus, and the man even showed him the best way to get there after his boat docked in Manaus.

It was still early in the afternoon, so Charley examined all the boats. He especially enjoyed all the fishing boats. As they came in with the day's catch, he checked out the many varieties of fish and creatures and especially enjoyed the happy chattering and good-natured jostling among the fishermen. Boy was it hot! As the position of the sun told him it was about 4:00 p.m., he went to the Amazon River Ferry Boats office nearby and bought a one-way ticket to Manaus for the next day. He spent a fitful night, partly caused by the heat, but mostly due to his excitement about the upcoming trip.

10

Charley Up the Amazon

CHARLEY WAS WAITING ON THE DOCK in the morning as the ferry pulled in. It would take at least a week to get to Manaus. You could get a cheaper rate by sleeping in a hammock than renting a cabin that had a bed, so Charley chose sleeping in a hammock on deck. The captain took one look at him and put him in a cabin. From that moment on, this old man carrying his small valise and his saddle wrapped with a rope got the run of the ferry and top-of-the-line treatment and food. The captain even spoke a little English.

The Amazon has very little fall in it, so it seemed like traveling on a lake. The Brasilians call Manaus "the city one thousand miles up the Amazon" [from Belem]. The boat was fairly slow, so it gave Charley a chance to see everything. The river in dry season is only six miles wide, but now, in the rainy season, it had grown so wide over the low banks that he couldn't see both sides.

The trees were filled with beautiful big macaw parrots. Their bright red and other colorful feathers shined out of the trees like Christmas decorations. There were uncountable numbers and species of swimming birds fishing along the river's edges. Large spectacular toucans were a common sight, perching around in the trees.

The jungle was so thick it seemed black. Sometimes the boat went near enough to shore that he could hear the beautiful wild sounds of the jungle. Flowers were blooming amongst the thickets and on some tall trees. Charley had always liked monkeys, and he thrilled at the sight of many wild ones, especially at the stops they made at small towns along the way. He imagined he could distinguish their wild cries from the other sounds. He had read of the Great Horned Frogs found only in the Brasilian jungle, and he saw one of these amazing creatures in a cage. He knew that a large species of jaguar roamed those jungles, but of course he didn't get to see one. Also lurking unseen, usually in the water with just their eyes above the surface, were the great anaconda snakes. They could squeeze a man to death in just seconds, and even large animals like cattle and horses were in danger from them.

Besides the snakes, an occasional crocodile further reminded him of the dangers of the waters of this river. The heavy rainfall made it all seem more ominous. He remembered the ravenous piranhas he had seen devour a whole cow and imagined there were even more of them in this water. Meanwhile, he had never seen so many mosquitoes and biting bugs in his life, and he wrapped up his body so only his eyes showed when on deck. The rats and cockroaches partied all night in his room, but they stayed off his bed. He rubbed tobacco juice on the legs of the bed so they wouldn't crawl up them.

Each time they stopped at a little town, Charley noticed that the natives were Indians, and they all looked like they came from different tribes. He supposed they must have been the descendants of the first people who inhabited this land. They were all happy looking and friendly to him. They showed him their fish and their fishing boats with great pride. He could only imagine what the fish must look like that they caught with a hook as big as a boat anchor. He wished he could have gone on a fishing trip with them.

Charley asked the captain what all those heavy-laden freighters they passed on the river were carrying. The captain said that it was mostly rubber and Brasil nuts, as well as rosewood oil and Jute, along with various forest products. He said the big barges going downstream were mostly loaded with oil from Peru going to the refinery at the mouth of the river.

It was November 28 by now, and it rained hard every day. Charley was really tired of being wet, his joints ached from inactivity, and his sinuses were swelling again, so he had a headache most of the time. The scenery by now had become monotonous in its sameness, and less wildlife showed up in the heavy rain. Finally, they arrived in Manaus.

Charley wrote home:

November 28, 1960
Manaus, Brasil
To: Mrs. Rex Walker
1300 Cherryvale
Boulder, Colorado U.S.A. It is postmarked Manaus, with Brasil stamps

Thanksgiving Day
Dear Dotter,
This is a kind of lonesome day. I got a room in a flophouse and it is raining like hell so I am just settin' here.

This is the place they call a Thousand Miles up the Amazon. Manaus – it is a big town – the Amazon is a really big creek. It is full of muddy water but the Rio Nigro is clear just about the color of tea. You can't imagine the bananas piled up on the docks – look like big hay stacks. Brasil nuts by the ship load.

There is a big English shipping company here that I would like to get some information on. I went down to the office but I couldn't understand the bastards any better than I can these Brasilians, but I met a Baptist missionary. He was real white man – he took me home for dinner – has a nice wife and five beautiful kids. The little one was born here and they named her Amazona. He was a different type than that bastard I met in Brasília.

I will leave here tomorrow and go up the river. I will write again when I can. Dad

Charley had been gone on this trip of his dreams now for five months, and his family could hardly believe that he was continuing on through the rainy season. Thankfully he had gotten all the shots required for visiting the jungle. And he was closer to home than when he first started exploring Brasil.

11

Charley Up the Pan-American Highway

TIRED OF THE RAIN FOREST AND THE TROPICS in general, Charley flew out of Manaus to Bogota, Colombia. It was a big, interesting city, and he spent several days there. He found a variety of people who spoke English, and picked their brains about everything. He eventually decided that there was nothing in Colombia he was interested in during this rainy season, so he flew from Bogota into Panama City.

He had always wanted to see the fabulous Panama Canal, so he took a good look at it. It was a wonderful thing! He marveled at the engineering genius of the builders of this passage from one massive ocean overland to another massive ocean. He talked to English-speaking people and discovered that, as he had hoped, there were ways to travel so he could see all the Central American countries on his way north. They warned him that the Pan-American Highway was far from finished and it would be a rough trip. The comforting thing was that he could get off a bus in any town he wanted to, and there would be another one along in a day or a week.

Tapachula, Mexico
December 28, 1960
Dear Dotter,
Well I am back in Mexico. This is a town on the Talisman River, which is the border between Guatemala and Mexico. [Even though there is a town named Talisman in Mexico, it is the Suchiate River that separates Mexico from Guatemala along its border with the state of Chiapas.]

> I flew from Colombia to Panama and then came up the so-called highway which isn't there but it was a nice trip. I enjoyed it a lot – seen some really good country all through Central America except Guatemala, I didn't like it.
>
> I wouldn't give one little part of Central America for all South America. It really is a grass country.
>
> I spent Christmas day in a little town in Guatemala about 9,000 feet up in the mountains. I damn near froze to death. Their houses are open just roof and floor and they don't have any stoves or heat of any kind – I don't know how them bare footed Indians stand it.
>
> I met a nigger in Guatemala City from New York. He told me they have had one of the worst storms in the States they have had for 28 years. I hope the boys wont have too much loss. I haven't decided what I will do now. I am thinking quite seriously of going to Australia, but haven't made up my mind yet. Think I will go to Mexico City and talk to Aussi. Ambassador, but I will settle down some place in the next few days and write you again. Dad

It was obvious from this letter that Charley was thinking about family and home—however, not yet ready to give up his dream of a better ranching country.

Charley took a bus from Tapachula to Mexico City. That monstrous city was a test of his ability, but he finally found what he was looking for. He made inquiries about Australia from the Australian ambassador in Mexico City. He got maps and air connections to cattle country.

Charley's hearing was deteriorating more. He found it difficult to carry on a conversation with anybody. He had always said he would never get a hearing aid, but now he changed his mind. There was a lot he wanted to do yet, and he couldn't do it without being able to hear.

Charley was done with South America and with Central America. He decided to go north into Sinaloa, Mexico, where he knew a little bit about the country from his trip there in 1950.

Charley wrote home:

Chihuahua
January 18, 1961
[Referring to a previous letter never received.] *I have written you before about Christmas but haven't got any answer. You haven't got it because you are so good about writing. I am fine, the weather nice and warm. I have fresh fruit and vegetables ever day.*

> *I am on the Rio Fuerte. It is just a little village, nobody here can speak English so I have to talk like they do. I am picking up quite a bit. There is a big hot spring here. I spend a lot of time in it. I really feel good. The people are sure nice to me.*
> *This is the address:*
> *Ferrocarill Chial Pasifica*
> *Agua Caliente de Lanpho*
> *Sinaloa, Mexico*

It was well into the month of January, 1961, by now. Charley took a bus to Los Mochis, Sinaloa, then got a ride about thirty miles out of town to the Imala's hot springs. He got a small room there, with a little restaurant nearby. The springs were not developed very much, but that didn't bother Charley. He soaked his aching body in the deliciously hot mineral water, slept, ate and rested. He got a lady to wash his clothes and his blanket and his saddle blanket. Once again he greased his saddle and his boots, which had really suffered from the time in the jungles. He saw he would have to buy a new cinch before he rode his saddle. The old one was totally rotten, and the latigos were doubtful. He needed some new clothes, but maybe these would hold together until he got to Montana, where he planned on going. He knew that Potch was going to be gathering some wild horses and he wanted to get in on it.

Charley sent a telegram. Queeda and Rex were now in Tyler, Texas, spending an extended Christmas vacation at Rex's parents' home:

Feb. 4, 1961
Telegram from Mazatlan,
Sinaloa, Mexico
To: Rex Walker
536 Park Heights Circle, TX
Just received your letter still in Mazatlan for several days stop Come if you can.
Charley Mantle

Charley sent the following letter to Tyler, Texas, and it was forwarded on to Boulder, Colorado:

Mailed February 10, 1961
Mazatlan, Sinaloa, Mexico
Queeda Dear,
Just got your letter. So sorry you couldn't come down. I don't

know why I didn't get your letter. I went to the post office every day. I got a letter from Potch that was wrote the 28th Got it the Feb. 1. Yours was wrote the 26 – didn't get it until the 4th.

I am leaving here tomorrow going over to Durango. Potch has a friend in Mont. that wants 3,000 steers and wants me to look around and see if I can find them. This is no cow country here. They just raise little mexs and donkeys.

I never have went fishing. I just don't feel up to it. I can't get rid of this damn cold. I thought this tequila would cure anything, but it just takes your toenails off and slips your hair. Potch said him and another fellow had bought 200 wild horses that has got me all hopped up – I want to get in on running them. Don't guess I could do much but I sure want to see it.

*Will write you again when I have something to write about.
Dad*

About March tenth he went back into Los Mochis and caught a bus to Mazatlan down the coast a ways, where he figured he would have to fly out of the airport there. In visiting with people around town, he came upon a Canadian man who had driven down to Los Mochis to do some dove hunting. He was ready to go home, and Charley made a deal with him to ride along as far as Glendive, Montana. Along the way, he called Evelyn at Queeda's house in Boulder and told her he was back from Brasil and headed up to Charley, Jr.'s, in Dickinson, North Dakota. Evelyn was so relieved, she just sat down and cried.

12

Evelyn, 1959 to 1961

AFTER CHARLEY LEFT FOR BRASIL, Evelyn spent the winter in Boulder with Queeda. She needed the rest, but she soon found out that this household was not a restful place. Queeda and Rex had a three year old, Cindy, and a toddler in the house. The grandkids were fun, but all of the other constant activity kept her worn out.

The toddler, Justin, inhaled a piece of a crayon into his lungs in October. He had to have surgery to remove it, and it caused him to have siege after siege of bronchitis all winter. Queeda was cooking and washing for a crew working on the ranch. Trucks bumped down the driveway day and night, terrifying Evelyn that a child would be run over. The phone rang constantly, and people streamed in and out of the house, usually with manure-laden feet. Evelyn seemed the only one who had time to spend with the children, and they nearly ran her to death. She was in the middle of all the hubbub, and she couldn't bear to just rest and watch, so she worked so much she became worn out. She also became very homesick for the peace and quiet of her ranch home. She worried and went sleepless many nights, thinking about Charley off alone in a strange land.

Tim had graduated from college in June 1960 and was immediately drafted into the army. His brother, Lonnie, had done his time in the army, and now took over the running of the Mantle Ranch from their brother, Pat, so Pat could get on with his life. Tim would get out of the army in 1962, and at that time he planned on returning to the ranch. Pat had gone into partnership in a rental horse business with his brother-in-law, Rex Walker. In the summer of 1959, they had launched the Sombrero Ranches, a business of renting riding horses to camps, outfitters, and stables. The headquarters was at 1300 Cherryvale Road in Boulder, Colorado, at Rex and Queeda's small ranch. Pat and Rex had also begun building a stable for their first location. As they built, they were renting out horses from this, their first Sombrero Stable. It was located on the east side of Estes Park, Colorado, right where US highway 34 burst out of the Big Thompson Canyon into the Estes Park Valley.

Evelyn got to be with Pat a great deal that winter. He was working to get contracts for his 7-11 Rodeo string for the next summer. He and Rex traveled to every town in Wyoming, South Dakota, and Nebraska that had a horse sale. They bought horses for the stables and for the rodeo string. Between sales they were building corrals. Buying hay for the summer took up a great deal of their time and money, and the rest went for repairs on the old trucks they had bought. Pat was a good mechanic and could do more with bailing wire than most men could accomplish with welding equipment. Horse traders made Cherryvale a regular stop to eat, sleep, repair their equipment, and sell a few of their horses to Sombrero Ranches.

※ ※ ※

Evelyn worried constantly about Lonnie down there by himself at the lonely, isolated Mantle Ranch that winter. She knew from long experience of the loneliness and hardships that came with that ranch. Unknown by her or the rest of the family, Lonnie spent quite a bit of time in Craig when ranch chores allowed. He had met a pretty girl named Grace Spykstra who was a schoolteacher in the Craig High School.

※ ※ ※

By March Evelyn was getting very anxious to get back to the ranch. She had the garden to plant, winter laundry to do, and a house to clean. She knew that Lonnie would be hungry for some home cooking, too.

When she got the call from Charley that he was back in the United States, Evelyn felt free to go. She was feeling much stronger again, but knew she must never again do all the backbreaking work she had done all her life. It was going to be hard to make herself back off from all she loved doing and was so good at. Her melancholy vanished with the arrival from New York of her dear first cousin, Satie Menon, and her husband, Mike. Satie had recently married Mike, and Evelyn had never met him. They seemed a perfect couple, and the three of them talked into the wee hours. Satie promised that they would come visit Evelyn at the ranch.

Back in Craig, Grace's students thought it was great fun to see who could be the first to spot Lonnie when he came into town. They would each try to be the first to tell their teacher the news. So when Lonnie came to town, she would know immediately, and she usually showed up where he was, saving him the trouble of finding her. Of course, the students kept pretty close tabs on the couple, and Lonnie and Grace didn't get much time alone when he came to town.

Lonnie's sister, Queeda, had gotten wind of Nav having a girlfriend. He was so tight-lipped about it she knew it must be true. Queeda and her two children, on a trip to the Mantle Ranch for a visit, went through Craig one day in May and found Lonnie sitting at the Midwest Café. She sat down beside him and they were visiting. The door opened and a dark-haired young woman came in with storm in her eyes. She headed directly for Lonnie. This had to be "the one," so Queeda put her arm possessively around Lonnie. Grace had never met Queeda, and this woman was sitting next to Lonnie. Grace approached the table and snorted something like, "who is this?!" Lonnie ducked his head, his face flaming red, and before he could say anything Queeda purred, "Honey, who is this hussie?" Lonnie just couldn't make his voice work, and next thing he knew Grace had stormed out of the café. She almost ran down the sidewalk, then all at once she stopped dead in her tracks. Whirling around Grace stomped back in the café to get her man. Queeda began angrily berating Lonnie for being unfaithful and leaving her and their two babies at home all alone. The entire café knew Queeda and were having the laugh of their lives. From the looks on the faces around her, Grace quickly caught on that something was rotten. Later, as in all small towns, the news really spread. Queeda feared for her life, but Grace was a forgiving person, and she and Queeda were the best of friends all their years.

Lonnie and Grace's courtship was progressing rapidly. Unofficially, they were engaged. He wanted to give Grace a nice honeymoon, but he had no money. He began cutting cedar posts and selling them. He sold truckload after truckload of cedar posts and firewood in Craig until he thought he had enough money to get married on. He wanted to make enough more to buy them a pickup, but Grace intervened. She had a nearly new station wagon, and she insisted that they drive it until it fell apart, and then they would buy a new truck.

※ ※ ※

Tim was discharged from the army early on a hardship application and returned to the ranch in the spring of 1961. He said that he was ready to take over the ranch, so Lonnie was free to go, or they could manage the ranch together. Lonnie and Grace set their wedding date for February 2, 1962, in Littleton, Colorado, where her parents lived.

When Evelyn got home in March, she found her garden plowed and ready to plant, all the ditches cleaned, the pump ready to drop in the river to water the garden, and the house as clean as a couple of bachelor boys could get it. There was even a big pile of chopped wood. She ran to check on her chickens and found them healthy and beginning to lay heavily. The milk cow had just freshened, so there would be plenty of milk, butter, cream, and cheese. She had never felt happier or more content to be home.

※ ※ ※

During that summer, Evelyn was very lonely. Lonnie and Tim were off tending the cattle and riding in local rodeos most of the time. Since the expansion of Dinosaur National Monument in 1960, it now engulfed all of the Mantle Ranch and its grazing permits. The Monument had taken over management of the grazing within its boundaries. Previously, the grazing permit had been under the management of Taylor Grazing, which is what the BLM operated under. Taylor Grazing was a good and fair system set up through a cooperative plan hammered out between ranchers and the government. When the National Park Service took over the grazing management, they knew nothing about it, micromanaging through use of inept personnel. They didn't know where the water holes were or even realize what a critical part water played in ranching and wildlife management in this dry, desolate country.

During their second year managing grazing in the Monument, the NPS changed the grazing permits so drastically that it cut the Mantle cattle herd by one-third. Yearling steers had to be taken off the ranch and pasture bought for them someplace until shipping time. This usually meant building fences on the pastures and creating a dependable water supply. It also was critical that they take frequent trips to the pastures to check on how the steers were doing. On the ranch, the new rules required a great deal of gathering of the cattle and moving them to different pastures, and destroying their natural flow to other parts of the ranch—all on the whim of the NPS. The only thing that was the same was that the cattle were wild and hard to handle and the land was just as rugged, full of box canyons and rocky rough trails as it had been for the last forty years of its existence. Evelyn's boys were only able to be at the ranch and see their mother during the time they were putting up hay. Carefully, they arranged their work so they could take in rodeos on the weekends. Not only was it a lot of fun, but was a source of income, too.

Evelyn had been terribly sick and had nearly died in 1958. She had had several setbacks since. She had moved her garden to a large fenced-in area below the house in 1950, but this summer the boys had prepared a small garden for her by the house that she could irrigate with a pump in the river. It was a real blessing, as she could tend it easily. There was still drinking water to haul, the trip to the river to gas up and start the pump, then the trip to shut it down, chickens to tend, the cow to milk with her rowdy calf to tie up while she milked, and the orchard to irrigate and pick and can. Friends came from time to time to visit and help her, but mostly she was alone.

She realized that Lonnie was in love, but she had never met the girl. She wondered just how in the world that would work out. She didn't think any female but herself would ever agree to live on this backward ranch. She made it a point to drive out to Craig one day with the express purpose of meeting Grace Spykstra. They met and had a nice visit, and Evelyn was well pleased with this girl. Out of the blue, Evelyn asked Grace, "Do you know what in the world Lonnie is doing with all the cedar posts he has been cutting?" Grace laughed and told her yes, she knew. He was selling posts and firewood in Craig to make enough money so they could get married and go on a honeymoon. Well satisfied with her trip, Evelyn went home happy.

Her children insisted that Evelyn spend the winter again in Boulder with Queeda. She agreed to it, as she knew everyone would just worry themselves sick about her if she didn't. As the time drew nearer, she got excited about going. Charley was not coming home, she guessed, so she didn't need to make a home for him.

13

Wild Horses

IN MARCH OF 1961, Charley was dropped off by the Canadian he had caught a ride with from Mazatlan, Mexico. He was only 100 miles from Glendive, Montana, where Charley, Jr. lived, and he picked him up. Charley, who always called Charley, Jr. "Potch," was so happy his eyes glowed at the thought of running wild horses. First of all, he bought some clothes: Levi's, a long-sleeved shirt, boots, underwear, socks, and a hat. He bought a warm windproof coat, a big, black silk scarf, and leather gloves. He found a place that sold the kind of chaps he liked, and he bought them and a heavy leather vest. He bought a new lariat, cinch, latigoes and saddle blankets, wrapped his saddle horn so he could hold a dally, in case he had to rope a wild horse, and he was ready to go chase wild horses.

Potch told him the lay of the land and all about the hundreds of horses running out there. A rancher named Jack Eaton owned the ranch they would be gathering wild horses on, [just outside of Glendive, Montana]. Jack was a throwback to the old tough cowboys of the West that history is made of. Jack had heard of what a hand Potch was with livestock, so he came to see him with a deal. He had bought up many small ranches in the area and ran his outfit under the <u>HOT</u> brand, and called it the Hot Bar Ranch. Jack hired Potch to cowboy for him. It was a good job and paid good money. Potch was overjoyed to be working on a cattle ranch once again.

Potch continued to work for Jack in the summer and fall. Jack's ranch was in an area of jumbled, rugged country known as "The Badlands." Many herds of wild horses roamed on this land. They kept the feed on the range picked clean, and Jack wanted the wild horses rounded up so he could have that range for his cattle. He asked Potch if he possibly would hire out to rid the range of these wild horses. Potch jumped at the chance to do what he loved best. He knew there were a lot of things he probably didn't know about trapping wild horses, but what he didn't know, Jack Eaton did—and just maybe he could get his father, Charley Mantle, to throw in with them, too. His dad had to be the best wild horse man still alive in America today.

The wild horse roundup project began in the fall of 1959. Potch had built probably the most magnificent wild horse trap ever known. The drive started on Glendive Creek where the wild horses came to water each day. When the horses came in from the rugged range to drink, the cowboys burst out of hiding and swooped down on them, yelling and cussing and scaring them into a terrified run to escape the attackers. The men on the fastest horses flanked the herd of wild horses, keeping them together, while the others rode behind, keeping them running so furiously they wouldn't recognize the trap they were entering. Impenetrable wings of brush and cable angled from a very wide entry and slowly diminished in breadth until only a narrow way led into the mouth of the box canyon. At the mouth of the canyon the cowboys jumped off their horses and pulled shut a mammoth pole gate, making the trap secure from the horses breaking back when they saw their mistake. Potch had built a fence along the top of the canyon wall to keep any of the trapped horses who might scramble up there from jumping out.

Marmion cleaned up the mouse-infested cabin at the old Foss place, where the headquarters for the roundup would be. She organized everything so she could cook for the cowboys. Once a week she went to town for groceries. Everybody had to sleep above ground level, as the place was infested with rattlesnakes. They found many skins the snakes had molted out of strung up in the rafters. Even though Marmion poisoned mice, stomped mice, trapped mice, and three-year-old daughter, Cammie's dog caught several a day, there were still mice. Luckily they couldn't get into the storage cabinet for the food. The shelves were covered by a tight-fitting door that let down on hinges to its fold-out leg to make a table.

She prepared meals on, and they ate at, this table. All her cooking pots were carefully hung on the walls.

Charley wrote:

> *I shore have enjoyed Marmie and Cammie this summer. Marmie is the sweetest girl in the world. She never complains. We are camping in an old cabin where an old bastard had lived for sixty years. I think it was the dirtiest place I ever saw. When we was cleaning it*

out we found magazines dated 1911. Rattle snakes I never seen so many in my life. We killed them on the door step, in the toilet, in the woodpile, saddle shed, everywhere, and good god amity man it has been 108 in the shade lots of days but Marmie never hollered. Cammie has attached herself to me like a wood tick. Ever place I go she is right there. She really plays up the Grandpa stuff. The other day her mother was going to spank her. She said, 'Mama I don't think Grampy wants you to spank his little girl!

From then on, everybody called Charley "Grandpa" and his son finally became "Charley."

Marmion would often ride out with the men when they were spying out a herd of horses. One of the men usually carried Cammie on his horse with him. When Cammie would get tired, Marmion would ride back home with her. Jack had an old saddle horse that was very gentle. This was Cammie's horse, and she led him around endlessly. When Marmion had time from her many duties, she would put Cammie on the horse and lead her around. Sometimes Cammie would insist on riding by herself, and then she and her horse would be put in the corral together. She seldom got tired of riding before she had to be removed howling and kicking from her horse.

Marmion's constant fear was for Cammie. This little three-year-old with the enchanting smile, big golden eyes, and silver-blond hair was everywhere. She was the darling of the camp. They had gotten a dog for Cammie to guard her from rattlesnakes. She and the dog went everywhere together. One day, nobody knew why, the dog snapped at Cammie and gave her a big angry cut on the face. Potch shot the dog, and they took its head to town to be tested for rabies. Cammie got many stitches but healed quickly. The tearing left only a small scar.

One morning Marmion headed for the outhouse. She just got into position to sit down when a rattlesnake rattled right beside her. She froze, thinking wildly about what to do. Calling on everything she had ever heard, she made herself stay perfectly still. After what seemed like an eternity the snake crawled off and disappeared through a crack in the wall. Marmion couldn't help herself, she ran gasping and crying out of the outhouse. Potch caught her and held her close while she told him what had happened. This fascinating, often-told story gave toddler Cammie her own snake story to top those the cowboys told while sitting around drinking coffee and spinning yarns about their snake encounters. The number of rattles lying on the table in the cabin was proof that most of the stories were true. It is best to not even mention snakes or mice to Cammie to this day.

Never one to allow peace in camp for long, Grandpa would put little pink baby mice in Cammie's pants pockets. She would proudly show them around. Cammie would also try to imitate the cowboys three-footing a wild horse. It was her dog she practiced on, which might account for the bite. She would tackle him and tie his feet together, just like the horses got tied. Grandpa had also captured a baby owl that became a pet. Marmion, who had brought out laying hens so they would have eggs to eat, hoped this horse gathering would be over before the owl got big enough to catch chickens.

From March to September, Grandpa spent running wild horses with Charley, Marmion, and Cammie. Each gather brought in about twenty-five horses. They were roped and sorted. The horses carrying brands were returned to their owners. The slicks (those without brands) were either sold, kept to be broke, or shipped out.

There was plenty of riding "hell-bent for leather." The saddle horses became lean and hard, tough as nails. Replacing shoes was a constant job. Grandpa wanted to help shoe, but Charley, Jr. wouldn't let him, as he didn't want his dad to be down in the back so he couldn't participate in all the fun of the wild horse roundup.

They would jump a bunch of wild horses, get behind them and as close to the sides of them as they could get to hold them in a herd. At breakneck speed they rode over rocky ridges, through thick thorns, slid down sharp inclines, jumped washes, and crashed through sagebrush. During all this, the cowboys let their imaginations dictate the terrible language and yells that came from them in a wild crazy stream. The wild horses were scared out of their wits and ran for their lives. If a wild horse tried to break out of the bunch and escape, presenting the chance for others to follow him, he was roped and tied down and would be brought in later on a lariat rope.

During one of the first gathers, Jack Eaton roped one of these escapees. The squealing, lunging half-choked wild horse ran off a bluff, dragging Jack and his saddle horse with it. The fall, and the fact that his horse fell on it, broke Jack's leg. Biting back his pain Jack finally got into a position where he could get out his pocket knife. Regretfully, he cut his lariat rope to set the wild horse free so he wouldn't choke to death.

Potch had seen Jack disappear over the bluff and raced quickly to his aid, dreading to find him crushed and likely dead beneath his horse. They got Jack loose from his horse. The horse staggered up onto his feet after he got his wind back, but Jack was not so lucky. It was obvious that his leg was broken. Potch found a couple of sagebrush trunks and made a splint from them, then boosted Jack up in the saddle. Potch hurried back to the herd so they wouldn't lose them, leaving Jack to ride painfully back to camp where Marmion got him off his horse, unsaddled the horse and turned it out, then she drove Jack to town to get a cast on his leg. Jack walked with a distinctive limp all the rest of his days.

Grandpa and Jack became very good friends. They had come from the same background and were both the best of the best cowboys. Both could tell a story that would leave their audience spellbound. Neither liked to eat their veggies and loved good horses and kids. Potch, Marmion, Cammie, who was three, and Charley all lived at what was known as the Foss place. Jack bought the Foss place from Old Man Foss, who was old and wanted out of the ranching business. In the past, Foss tried to keep the wild horse numbers down, but as he became older and the cowboys able to do a roundup of wild horses became an extinct breed, the wild horses had overrun his ranch.

Foss's arch enemy became a little stud that was just too smart to catch. He was called "The White-Tailed Stud," and he became a legend in Montana. He was a rather small horse, all black except for his white mane and tail. He escaped with some trick on the riders every time they had him at the point of no return. After he broke back to freedom, he would race out to a high hill where he would prance and snort and squeal his victory and challenge to the crestfallen cowboys. During the big, wild horse roundup, he tricked them many times, but finally one day his luck ran out and he was captured. Cowboys respect a horse like that, and they did him no harm. Jack kept him and broke him, and they rode hundreds of miles together. A beautiful painting of that horse hangs on the dining room wall at the Eaton's ranch.

On September 4, 1961, Charley wrote:

We started on these horses in March and been getting up at four and five o'clock and get back to camp from eight to midnight. Right now we are takin a little rest. The saddle horses are so skinny and sore footed we had to quit for a while. The folks have gone to a rodeo some place they call State Line up towards the Canada line.

Never have heard if Tim got out of the army or not. He wrote and wanted me to make out some papers. I was up here (at the Foss place camp) and had no way to get to town so it was quite a while before I got in. When I did it was in the night. Everything was closed up. I asked a feller where I could find a lawyer. He took me to an old shack. We woke a feller up – I told him what I wanted him to write and I sent it to Tim. Next time I went to town I saw that lawyer sweeping out the saloon. Was glad to see me and wanted to know if I had any more business for him.

I don't think we can work much longer. The fall rains are settin in and I think winter starts pretty early here. We took a few days off and went over on the reservation and helped Joe (Chase) brand. That is the best country I ever saw. Lots of grass and water. I am going back over there. Joe and his brother got lots of cattle. Charley and Joe want to buy 200 yearling heifers. I wish you would find out for me if the Ekkers have had a dry summer down in Robbers Roost Country. They might want to sell some cattle. When we leave here going to go help the boys wean. By this time it will be getting cold and I am going to head for Australia.

I wish I had someone to go with me. I am so deaf I don't understand hardly anything I hear. Well it is time to wrangle and get a night horse. Love, Dad

※ ※ ※

When the family moved back to Glendive that fall, Grandpa kept very busy all the time. He made jerky so he could be sure nobody went hungry. He found a hair twister and set it up on a post and made a lot of horsehair ropes. He cut up beef hides and braided hackamores from the rawhide. Every rancher wanted his handiwork, but he never sold any. He gave a lot away, though.

When fall came in 1961, Charley got to feeling the wanderlust again. He still wanted to go to Australia. He had heard that you could own as far as you could see in all directions, and it was wild and untamed land. He didn't want to see the "city part" around Sydney; he wanted to go north into the ranching country. He didn't want to see no damn sheep country, neither! He talked about it and begged him to go until Potch finally agreed to go with him. Charley would pay the way. Marmion was pregnant, but Potch supposed they would be back in time for the birth of the baby.

Their first obstacle was obtaining a passport for Potch. During the winter of 1961, Marmion, Grandpa, and Potch went to Portland, Oregon. Marmion's family lived in Portland and they were thrilled to see them, especially granddaughter Cammie. But obtaining a passport was no easy task, and they had a lot of trouble finding someone to help them. In the end it was a woman who worked at the travel agency where they bought their tickets that found a way to get Potch a passport.

Since they could leave for Australia right away, Marmion took Cammie and drove home alone back to Glendive. She just hoped it was a mild winter, since she didn't have any help.

14

Lonnie Moves to Wyoming

COME JANUARY 20, 1962, there was a great gathering of the Mantle family. Grace Spykstra and Lonnie Mantle were getting married. Everybody looked so beautiful and so handsome in their pretty dresses and dress suits. Grace's sister, Ann, was her maid of honor. Her other sister, Carol, was by her side. Her parents, uncles, and many friends gathered in. Lonnie's brother-in-law, Rex Walker, was the best man. Lonnie's older brother, Pat, and his younger brother, Tim, stood proudly by him. The reception was held at Grace's family home near Littleton, Colorado. The house was so full it seemed it might burst with all the merrymakers. Food just kept coming, and everybody was just a little sad when the couple changed their clothes, were bombarded with rice, and got in their car to leave.

Lonnie told Grace she should drive, and they headed out for their honeymoon in Grace's station wagon. Or did they? The car started right up, purring like a kitten. Grace stepped on the gas, and nothing happened. The motor roared, but the car wouldn't move! Suspicious by now, Lonnie got out and looked the car over. Besides every sort of rattling trash available being tied to the car, the back two wheels were sitting up on blocks. After a lot of good-natured torture, they were finally set free and set off on their honeymoon.

Tim was the only one who knew that they were not headed back to the Mantle Ranch now or ever. Lonnie had "diplomatically" told Tim that he and Grace were getting married and that he was leaving the ranch. Lonnie also told Tim that as bad as he hated to do it to him, he was giving him this pile of rocks they called a ranch. Given the grief that the National Park Service dealt the Mantle family over the next forty years, Lonnie's statement was somewhat prophetic.

Lonnie and Grace moved to Riverton, Wyoming, in mid-1962. Lonnie's close college friend, Jim Karman, lived there, and that was mostly responsible for them choosing Wyoming. The first year, Lonnie and Grace spent most of the summer going to rodeos. Lonnie participated in amateur rodeos. He entered the saddle bronc, bareback bronc, and bull riding events. The winnings got them happily on down the road to the next rodeo. He more often than not placed in the events, and his name became well known. Even the local meat packer where Lonnie and Grace purchased their beef told them to go to the rodeo and win enough money to pay for the beef. They pretty much lived like gypsies, surviving out of the back of their station wagon. Evelyn gave them an electric frying pan, so when they stopped any place with electricity, they could have a hot meal.

They first rented a house in Riverton, Wyoming. Neither Grace or Lonnie wanted to live in town, so they found a house that would do and when it came available, they rented it. It was just outside of Riverton on the Arapahoe Indian Reservation. The house only had electricity, no running water or plumbing. In the spring, a flood came through and during that short time, they had running water—so much running water they couldn't drive out to the road for the mud.

Lonnie began shoeing a few horses for the neighbors and people he rodeoed with. Nearly everybody in the area had a horse, and when they heard about Lonnie, they would call him to shoe their horses. He had such a thriving business that he finally had to start turning people away. One day when his friend, Jim Karman, was out visiting him, they got to talking about this problem. There had to be a better way to shoe a horse than breaking your back as you slowly went from foot to foot. Shaping the hoof, shaping the shoe on the forge, then nailing it on and clenching it and rasping it. Throughout this, you got little jerks at best and an all-out war with the horse at worst. Some horses even had to be tied down before it was possible to get the shoes on them. All of this put incredible strain on the back. A man could shoe only a very limited number of horses in a day. Lonnie and Jim tossed around a few ideas, then parted with the problem unsolved. Over the next three years, they would come up with a plan that would revolutionize horseshoeing for Lonnie and many other ferriers.

They had been in Wyoming long enough now that they found they liked the country and the people and wanted to stay there. Lonnie started looking for a place to buy, with some land and a better house. In late 1962, they bought seven-tenths of an acre with a modern small house on it between Riverton and Hudson, Wyoming. With this land Lonnie could bring some of his cows out of Colorado and lease pasture for them on the Indian reservation.

The leased land had to have signatures of approval every five years from the Indian heirs, who owned the property. One of the leases had 160 signatures to get. It took a lot of time to track down and get the signatures of all these people. It was a big reservation, and people lived in every imaginable type of home. Some didn't have a road to their house, others were leary about signing what they couldn't read. Lonnie was able to get most of the signatures, but every once in a while Grace would take an Indian lady friend who knew where most all the heirs lived, and spend a day driving around on the reservation getting signatures. This was always quite an experience. Sometimes the Indians were in bed, other times you had to take them to town before they would sign. Others were just plain drunk and wanted money for another beer. They got all the land they needed leased, and soon the Indians got used to the Mantles and became neighborly. Some would come by to visit, and there were always friendly waves when Lonnie drove by their houses or met them in town.

Lonnie still rode in the local rodeos, but even though he won consistently, they didn't pay enough to make much more than expenses. There were a whole lot more expenses now, as on May 4, 1963, Kail Everett Mantle was born. A perfect little boy came howling into the world, and their life was changed forever. Grace had asked Evelyn to come, but at the last minute, Grace's mother, Nellie, was able to be with her for the birth of their first child, so Evelyn did not go.

About this time, Lonnie and Jim Karman turned out their first shoeing chute. It consisted of a narrow box with solid doors in both ends. The box was made of strong timbers butted close together so a horse's hoof could not slip through a crack. It was completely lined with thick padding. A horse was led into this narrow box and both ends closed. The horse was strapped firmly into the interior, then a hand-cranked mechanism turned him, box and all, on his side, with all four feet sticking out the bottom. With padding and straps, the horse was immobilized and his feet secured. From that position, a team of two men shaped his foot, then shaped the shoe on a forge, and nailed the shoe on. The nails were crimped and the shoes made straight and tight before they were finished. This took a very short time, then the horse was cranked back to a standing position and allowed to catch its balance as all the straps were removed, leaving him standing in the box with only a halter. The chute door in front of the horse was opened and he walked out.

Skeptical people brought their horses to be shod and watched the box-shoeing with fear and dread for the safety of their horses. It was also pretty generally believed that the shoes would be put on crooked when done at this angle. Shape the foot with a grinder? Lonnie must be crazy! Their first time in the box, the horses were afraid, but after that they mostly took it in stride with no great fuss. Some old-timers just wouldn't accept it, so Lonnie still shod their horses on the ground. Any that fought him, he refused to do unless their owners let him put them in the chute. During its early days, there was usually a big audience around to see this contraption work. As time went on Lonnie altered, added, and changed the shoeing chute's features until it functioned perfectly. He and one assistant could shoe thirty horses in a day.

15

Australia, 1962

UNAWARE THAT LONNIE WAS GETTING MARRIED, and intent on their travel plan, the adventurers—Potch and Charley—flew out of Portland, Oregon, in January 1962, bound for Sydney, Australia. From Sydney they immediately changed planes and hurried on. Their destination was the Northern Territory near the town of Catherine, where Charley's research told him the big cattle ranches were.

They ended up near the equator, so it was hot, but the countryside was a far cry from the rain forest Charley had encountered on the Amazon River near the equator. It was a dry, parched, dusty land that looked like only a horned toad could live on it. Potch and Charley stopped in at the headquarters of a big cattle ranch they had been told about in town. Interested in these two American cowboys carrying their own saddles, the foreman invited them to go with his outfit on a big muster (gather) with some Australian stockmen in the area. This was a yearly event where they gathered the cows and calves off the range to sell.

Potch and Charley glanced at each other with an "I don't believe it" smirk on their faces over the saddles the cowboys rode. They had no saddle horns and free-swinging stirrups. They were English saddles, for crying out loud! With English bits in the bridles! The country they rode into was desert dry with very little grass, and almost barren of vegetation except for heavy thickets here and there. The brush in there was so thick and full of thorns it was impenetrable. There would be no gathering anything out of that brush. It would tear up your horse, your equipment, and even yourself.

A river ran through the ranch land. The plan was very simple, or so it seemed. The riders would post themselves where trails led off the range to the river, then round up the cattle when they came in for a drink. They would trail them out of sight and hold them in a herd while some of the riders went after more. Charley and Potch marveled at what excellent riders these men were on their "toy" saddles. With only a gentle English bit, they could get their horses to do anything they wanted them to, and no matter how wild the ride, they were never out of control.

The cattle were wild. It was obvious that some had never been gathered before. The Aussies could handily control the ones that just ran, but those that hid in the thorn brush thickets were a huge problem. When riders spotted a calf they would send in their cow dogs, who would chase the calf out. Usually a furious cow would come bellering out of the brush to punish the dog that was chasing her calf. All the Aussies were able to do at that point was to gang up on the cow and chase her until she and her calf ran into the herd. They had no saddle horns on their saddles, so roping the cow was pointless, because they couldn't control the animal even if they were able to catch one.

Potch and Charley watched awhile, then shook out their loops and took out after one of the escaping cows. Potch caught the head and Charley got the hind feet. They held the cow stretched out helpless until she quieted down. Then, when they let her go, the herders were able to easily drive her and her calf to the herd. Many times they did this, until their horses were tired out. The Australian cowboys grinned and cheered. One man asked Charley what they ought to do with those wild bulls that were hiding in the thorny thickets. Charley told them to shoot the bastards and leave them for the buzzards. He had already spotted the rifles strapped onto each saddle, so he figured that was what they did.

Potch and Charley were a huge attraction around the ranch, especially at supper time when the stories were told. They told real stories and tall tales and everybody drank up and howled with laughter until tears ran down their faces. They were careful to give Charley a gentle horse every day. Potch was a different story. When they found out he could ride 'most anything they gave him, it became a guaranteed rodeo every morning and noon when they got fresh horses. Australians obviously liked to bet, even if it was just on how far a kangaroo could jump, so wages changed hands before muster daily with bets on Potch or the horse. The foreman tried to hire the two of them. The American cowboys said thanks, but they had to be on their way.

Potch flew home in April, as their baby was due in May. Charley stayed in Australia another six weeks. A guy he met wanted to take him on a hunt for water buffalo. Potch was glad that he got to see

Australia, but he told Marmion, "We thought we had it tough in the canyon ranching, but that's nothing compared to ranching in Australia." Potch and Charley left their saddles over there. After the damage from brush and bugs, there wasn't much left of them.

Marmion wrote to Evelyn that Potch had made some plans, making it uncertain if he would be able to take as good care of her, the baby and Cammie as would be needed. Evelyn was happy to go to help her through the difficult first week. She arrived in Dickinson and Potch met her at the bus station. He drove the curvy rode back home like a New York taxi driver, and much to Evelyn's embarrassment, she got car-sick and they had to stop. The ride continued at a more comfortable speed.

Evelyn and Marmion bustled around getting ready for the baby, and Evelyn drove Marmion to the hospital in Dickinson one night soon after her arrival. Chari Loy Mantle was born on May 19, 1962. She was a healthy, beautiful brunette from the first and the darling of the whole family.

Evelyn got a surprise phone call one day from Rex's parents, Ross and Freda. They had been in Colorado to visit family, and then drove north into Canada. They drove west through Canada to Seattle for the World's Fair. They were very excited, as they did not often take trips. When they got there, they were disappointed at the impossibly long lines, but they enjoyed the fabulous new creations they were able to see. They gave Evelyn all the latest news from Boulder about Queeda and the children. They had already moved to Estes Park to the house they rented there. Queeda had come up to take care of business at the stable for the Memorial Day weekend, and they were resting up and playing now. Three-and-a-half-year-old Cindy and two-year-old Justin had caught fourteen of the little mountain trout from the stream behind their house. Grandpa Ross was amused that they had turned over cow patties to find fishing worms when they ran out. It reminded him of his childhood days in Louisiana.

Evelyn stayed a little over a week with Marmion after she got home from the hospital with baby Chari, insisting that Marmion spend most of her time in bed. Evelyn prepared good, hot healthy meals for her and Cammie. She scrubbed and dusted every corner of the house because Marmion was an immaculate housekeeper, and Evelyn didn't want her up and cleaning right away. Cammie loved to help with her little sister and was sweet and gentle with her.

Then Evelyn announced that she had to get back to the ranch.

16

Never Say Retire, 1962

EVELYN FELT GOOD ABOUT MARMION and her two girls. Marmion was up and getting stronger daily after the birth of her baby. She would be able to tend the children without Evelyn's help. Potch was back in the country, so if she were not here, Evelyn felt he would take over more of the daily care of his family. She was nervous about getting back to the ranch. Tim would be needing help. He had called her once when he had been out to town. He told her it had been a terribly windy spring, drying out everything, and the tourists were thick on the river and on the road. The ranch could be vandalized if nobody was there. She knew that Tim would not be able to take care of the home place, because he would be gathering the steers and putting them on a pasture he had yet to find, probably around Steamboat Springs.

Evelyn returned from North Dakota through Riverton to see Lonnie and Grace. They brought her home to Artesia. What a trip that was! On the way, they stopped in Rangely, where Lonnie had his entry in at the rodeo. Tim was there, and also their good friend Dick Macintyre. Between the three of them they won all the riding events, and Lonnie won the all-around cowboy title. Evelyn screamed and cheered until she lost her voice. After the rodeo, Lonnie and Grace had to go back to Wyoming. Grace was working as a teller in a bank back in Riverton, so they could only spend the weekend. Evelyn drove her blue pickup home to the ranch from Artesia.

Evelyn arrived back at the ranch on June first. During the long hot summer, she saw Tim only when he came down from Summer Camp, where the cattle were pastured. He came to the ranch only to put up the hay when it was ready to cut. She was very happy when Charley's sister, Nancy, came to stay with her for a while. They had been friends through some very tough times since Evelyn and Charley had married. Nancy never complained, but she was getting pretty crippled up. One of her legs had drawn up, so she was pulled forward into a stoop and was in great pain. She worried constantly that she was a burden on Evelyn, when in fact she was most welcome and even needed. In July, Nancy moved out to a room she rented at a motel. Life was just too difficult at the ranch in her crippled condition.

It seemed that Evelyn was surrounded by death and pain. She got word that their old friend, Shorty Chambers, had died. She wrote a letter telling Charley and sent it out to the post office with the first person who came along. One day Evelyn had trouble lighting her Coleman gas stove. Finally it blew up on her and badly burned her hand. How it hurt! She slept fitfully at night, and it was a daily struggle to keep the raw, painful burn from getting infected. She tried to tend the garden and keep the yard alive, but everything she did was painful. Just washing and preparing the vegetables from her garden to eat was a dreadful ordeal.

※ ※ ※

One day a Dinosaur Monument pickup came racing to the house, followed by a long trail of dust. The excited ranger told Evelyn that a boy was missing and presumed drowned in the Yampa River near the ranch. He had been on a boat trip. They needed permission to launch rescue efforts from the Mantle Ranch. It had taken them a very long time to get to the ranch over the road, which was in terrible repair. This delay caused the rescue to begin much too late. It took them eleven days to find the body. Apparently humiliated by their now-exposed failure to maintain the road properly, the Park Service took steps to drastically improve the road to usable condition. That, at least, made Evelyn's life somewhat easier.

※ ※ ※

By July, Tim was ready to gather cattle and drive them to the summer pasture on Blue Mountain. Unexpectedly, help arrived. It was Rex and Pat's attorney, Stan Johnson, with two of his friends. They had come from Boulder to be cowboys on this yearly Mantle Ranch shoveup. Stan brought his wife, Pauline and their two children—Jennifer, three years old, and David, three months old. One friend was

Dan Nolan, an insurance adjuster by trade, recently in from the wild western state of New Jersey, with his wife, Jean, and three kids, Bobby, Chris, and Steve. The other friend was George Nistico, an aluminum storm door maker and intermittent restauranteur from the extensive livestock ranches of New York City, with his wife, Elaine. The men brought their saddles and paraphernalia for the job at hand, and the women and kids were going to stay with Evelyn and help her cook for the shoveup.

Tim drove in with his saddle horses in a trailer and a ranch hand named Jess Lancaster with him. He had extra horses for his three new cowboys. Stan had gotten in touch with him that he would be coming and bringing three hands to help him. All Tim's apprehensions were stilled when he looked over his crew in their cowboy hats with tightly rolled brims, blue jeans that hit halfway up their boot legs, and wild-colored scarves tied around their necks. All three walked with distinctly bowed legs. He barely stifled a belly laugh when a New-Yorker-flavored drawl came out of two of them. Over supper, the good humor and excitement replaced any pretenses and apprehensions. Tim was a very good storyteller, as was his father, who they all knew, and they laughed until tears rolled down their faces.

The next day they all saddled up. Tim checked all the saddles, they kissed the little women good-by and rode off to round up them doggies. They rode out of Red Rock Canyon and started with gathering on Red Rock Bench. An old cow would raise her head up out of the sagebrush, lift her tail high in the air, and take off for the cedar trees. Tim would yell, "Stan, get around her and head her east. Bring all you see with her." A little later he would yell, "Dan, don't chase her, ride out around to the front and turn her." Dan and George had never ridden a horse at full gallop chasing anything before, let alone trying to control the horse, hold onto the horn, and give any kind of guidance to the horse while it jumped over sagebrush and lunged and dodged, feeling for all the world like it was bucking.

They got the cattle gathered and began herding them east. Stan was the most experienced cowboy, so Tim put him behind the cattle to keep them moving slowly east while Tim gathered more and threw them in with the bunch. Dan and George spent the day amongst the cattle, sometimes up front, sometimes in the midst of them, but always holding the saddle horn, leaning back, yelling "Whoa!" When they got the cattle to Hells Canyon it was nearly dark, so Tim shut the gate in the fence that would keep the cattle from returning the way they came and going back to Red Rock. He said, "Well, men, we did a helluva job today. Let's call er' quits and go eat supper and hit it again tomorrow." To his crew's horror, Tim took off in a long, jarring trot down the canyon for the two miles home.

They unsaddled and mumbled answers to the excited queries of the families about the day. Then Tim said, "Just time enough for a quick dip in the river. The cold water sure is good for saddle sores and sore muscles." His crew couldn't get to the river fast enough. They all took a long soak in the icy water.

The home crew had baked bread, picked cherries off the trees, and made cherry pies, dug potatoes for supper, and cooked several kinds of vegetables from Evelyn's garden. They had also milked the cow and made butter. How those cowhands could eat! They laid around on the lawn after supper and Tim began telling tales. All at once he looked around and every man, woman, and child was asleep.

Before daylight next morning, a great and terrible bellering yell from Tim yanked everybody out of bed. Blindly they pulled on their clothes over aching bodies and oozing sores. There was great moaning and groaning, but everybody drank lots of coffee and ate huge plates full of pancakes, eggs and bacon for breakfast. Wiser to the ways of the West than they had been yesterday, each man tucked a couple of pancakes in his pocket for lunch. Hitchin' up their britches and pullin' down their hats, the crew gathered round to hear the plans for the day. Tim told them to saddle up and load their horses in the trailer, as they would be trucking them several miles to start the gather today. All of them hurriedly curried their horse's backs and bellies where the cinches and saddle would rub and checked their horses for sores. The horses were better off than the men were, so they saddled them up and loaded them in the trailer.

The old pickup labored up the steep hill out of Hells Canyon, then bounced and lunged up the rough road. About 20 miles up the road, they jerked to a stop and Tim told the men to unload their horses and mount up. Jess took Stan with him and took off at a lope for some cedar breaks off to the north. Tim took his other two men and headed off to the east. Soon they were out of sight of each other or any other living thing. They would occasionally hear popping, thrashing brush and catch a glimpse of a red blur that must be a cow headed away from them fast.

Stan suddenly arrived at the end of the world. His horse stopped at the rim of the Yampa River. He was a thousand feet above the river, which wound its way west between the canyon walls. It was a sheer drop to the river. The man sitting on his horse beside him told him that the mountain across the abyss was Douglas Mountain, and the drop was just as high from that side to the river. Amazed and humbled, Stan Johnson would never forget that sight. In the many years ahead, he would get to know that river and this land well as he defended the rights in Federal Court of the Mantle family on the Mantle Ranch. "Back to business," Jess said, but Stan wondered just where were they going to ever find any cattle to herd in that jumble of high cliffs, cedar trees, and dark shadows that lay around them.

Soon there was a popping in the brush nearby, and Jess took off racing into the cedar trees, with Stan's horse close behind. There, in front of them, were five head of cattle, running like the hounds of hell pursued them. Jess motioned for Stan to stay behind them, and Jess took out to head them off. Finally they turned, but not before Stan had been nearly beaten to death and swiped off his horse by cedar limbs as his horse crashed after the cattle. They drove the cattle out into a sagebrush flat, and Jess told Stan he was not to drive them, especially, just—keep them in sight and let them settle down while Jess gathered more. Stan heard some yelling he recognized as cattle herding sounds off in the cedars, so he figured the other guys were coming with a bunch. Good, because he had told his friends they would be driving a great herd of cattle to summer range. Jess and Tim brought in the stragglers, and soon cows were bawling and calves were frantically answering as they paired up in the flat. As soon as the pairing up of the mothers and their calves was finished, Tim told the men to move them out.

Tim told the three new cowboys to bring up the rear, driving the cattle as slowly as possible, but pushing the laggards to keep the herd together. Meantime, Tim and Jess each took a side and kept the herd together and pointed west. Every once in a while, one of the cowboys would jump up some more cattle and throw them in with the herd. It was amazing how docile the cattle acted in a herd, considering how wild they had been when first jumped up. Little did the greenhorns know that there were cattle out there still who would have to be run down and often even roped to gather them.

The five cowboys slowly drove their herd down the long trail toward Hells Canyon. At the east rim of Hells Canyon, they turned the herd toward the foot of Blue Mountain, which loomed above them. They rounded up the cattle from yesterday's gather and threw them in with the new herd. Eagerly, the cows—who knew they would soon be on cool mountain pastures—found the trail and headed up the mountain. The calves were tired, their tongues were hanging out, dripping with sweat, and they wanted to hide and lay down. The hands must guard against the danger of some sneaking off into the trees and hiding. The cowboys really had to keep up their vigilance. That entailed some riding off the trail on the side of that incredibly steep, rocky, tree-choked mountainside, where only the brave dared go.

Stan all at once found himself slowly slipping off his horse, saddle and all. He managed to free himself from the saddle, but found himself standing on a shifting gray shale slope. He knew if he started falling he could never stop. Through great effort, he still had the bridle reins in his hand, and that steadied him enough to keep from falling. There horse and man stood, together, maybe destined to start sliding and tumble into the depths of Hells Canyon yawning darkly below them. The horse was trembling but stood still as a statue. Stan was able to unfasten the cinch and the flank cinch and lift the saddle from under the horse and up onto its back. Together they struggled back up to the trail, where Stan mounted up. The cattle moved slowly on up the mountain.

Finally, they got the herd to a flat they could easily guard so the cattle could not return down the mountain. They rested the herd while the mothers found their calves and let them suck. Separated cows and calves always return to where the calf last sucked, so they must be paired up after moving to a new location before leaving them.

The cattle moved along nicely now, as they knew where they were going and were eager to get there. The tired cowboys returned to the ranch, all except Jess, who rode on with the cattle. The soak in the river took a little longer than the night before. The new ranch cooks had prepared a wonderful meal again. With bellies full, the suffering cowboys piled into bed.

The banshee yell once more ripped through the early morning darkness, and the groaning cowboys pulled on their clothes. The motley crew with three days' growth of dirty beards and dust-caked hair, torn clothes, and sweat-stained hats gulped down the healing coffee and ate a big breakfast of pancakes, bacon, and eggs. Tim told them that today there would be no riding. Yippee!!! They were to get in the pickup and they would drive to Summer Camp on Blue Mountain, where they would brand the calves they had gathered yesterday. Piece of cake!!

The cattle were already gathered into a big round corral when they got to Summer Camp. Jess and another mounted cowboy were cutting out the cows and putting them out the gate, leaving mostly calves in the corral. A half-barrel of belching fire was heating branding irons just outside the corral, with the branding iron handles within reach from inside the corral. Beside it was vaccination equipment and an empty bucket.

Tim said it was time to begin, so they took a big drink of water, pulled down their hats, and they were ready to go to work. He told them to stand outside the corral, and when Jess caught a calf, to run out and grab it and throw it down and hold it down. Got it! Jess caught a calf around the neck, dallied the lasso to the horn of his saddle, and dragged the calf toward the center of the corral. The three new hands tackled the calf and tried to wrestle it down, then they tried grabbing its feet and tipping it over, while all this time the calf was using them for a battering target.

Tim yelled, "Stop!" He said he would show them how to do it. He walked down along the tight rope, letting it pass through his hand until he got to the calf's head. He quickly bent over the calf, grabbed him by a front leg and the flank, and when the calf jumped, he easily pulled him over onto his side. Then he jerked a short rope out of his back pocket and tied a front foot and two hind feet of the calf together. He lingered by the calf until he was sure it would stay down, then stood up. He told Stan to run and get a hot iron. Stan was glad to see a glove there, so he grabbed the handle of the branding iron with the glove and ran back to the calf. Tim took the iron and pressed it to the ribs of the calf, where it sizzled and smoked. Next, he showed the men how to vaccinate the calf. Then he took out his pocket knife and castrated the calf. He handed the testicles to Dan and told him to put them in the bucket. Lesson over, it was time to let the calf up.

It was a long, hot, hard day's work. The three now very experienced cowhands were bloody, filthy, exhausted, and hungry. Tim led the men to the little log cabin on the hill nearby, fired up the cookstove, and told the men to get cleaned up and supper would soon be ready. They went outside and Jess showed them where the spring was. Down the hill about fifty feet from the cabin, there was a bucket on a rope by a well. They pulled up bucket after bucket of the cold clear water and washed the sweat and grime and blood from their hands and faces. Their stinking clothes would just have to wait. When they got back to the cabin, Tim had laid out a feast. He had baked big delicious sourdough biscuits, fried up a huge pan full of meat, and made milk gravy from the drippings. He also had fried up a big skillet of potatoes. They dove in and devoured the food. Finally, George asked Tim what that delicious meat was. He said, "Calf nuts." His cowboys recovered quickly from the revelation and agreed that they were the best they ever ate.

Tim drove the exhausted men back to the ranch and they fell gratefully into bed. The next morning they had to drive back to Boulder and the real world. They would never forget this big dose of the West they had gotten. The women felt that they had learned a whole lot about ranch life from a real expert. Evelyn was their hero! She had thoroughly enjoyed their company and their help. With her burned hand and the pain it caused her, she would not have been able to cook for this crew without their help.

※ ※ ※

The lonely summer wore on for Evelyn. Her hand finally improved past the critical stage. One day she saw a big black Cadillac rolling up to the front of the house. Out stepped Ross and Freda Walker and Alva Jean, the young black woman who worked for them. Best of all, they had Cindy and Justin with them. What a joy they were to see! The two children were bouncing with excitement.

Ross, Freda, and Evelyn had a wonderful time sharing the events of their lives since seeing each other last. They had become fast friends from the time of the marriage of their children. They picked fruit from the orchard and fresh vegetables from the garden and watched the milking of the cow. They spent what seemed like hours sitting on the riverbank, watching Cindy and Justin wade and splash. Alva Jean was terrified all the time she was there. It was just too quiet, the towering cliffs silent and ominous, and the night sounds were strange to her. She had never seen so much darkness or so many stars in the quiet skies. She wanted very much to be back in old familiar Texas.

They had come with the intention of taking Evelyn back to Boulder with them, but Evelyn just couldn't go. There was altogether too much to do, and she felt that Tim needed her to take care of the home place. They got her to agree that she would come to Boulder again for the winter. Everybody was worried about her, down here all alone.

※ ※ ※

Evelyn went to town on October 22. She had gotten some supplies for the home place and had taken hunting camp supplies to the Summer Camp. She left a horse in the corral so that upon her return, she could go get her milk cow, which she knew would have wandered far down the river. While in town on her trip, Evelyn found Aunt Nancy living in a cold apartment in the tiny town of Blue Mountain. In her crippled up condition, she couldn't take care of herself, so Evelyn brought her home with her. She knew she couldn't get Nancy to stay very long, but she would care for her for a while. Evelyn wrote to Queeda that she would try to be in Boulder by Christmas, but things were uncertain.

The big news was that Tim was getting married. The wedding date had been set and it would be in Boulder. He met LaRue Betke in Boulder. She was an airline stewardess for Western Airlines. She was originally from Nebraska. Tim met her at the Green Meadows Ranch, a riding facility in Boulder where Pat and the ranch owner, Buddy Hays, put on a little rodeo every weekend. Tim had fallen for her hard. He brought her to the ranch and showed her around, so she would know what she was getting into. She liked what she saw, and they planned on spending the winter at the ranch.

The news came in that bear hunters had scattered the steers from the pasture at Steamboat. The poor wild steers, unfamiliar with the country, had scattered everywhere. It was going to be a very grueling roundup, if not an impossible one. Pat came over to help Tim.

Queeda, Cindy, and Justin drove in one day for a short visit at the Mantle Ranch. Business was over at the stable for the summer and they had a few days free. Evelyn basked in the luxury of having her daughter with her and hugged and coddled the two kids. They remembered all the fun swimming last summer and insisted on going swimming in the river. It was getting very cold, and the goose bumps raised up all over their little bodies, but they laughed and splashed and screamed at the cliffs to hear the echo. They had to be bribed with apple pie to get out.

※ ※ ※

Pat had a gut-wrenching decision to make. His old friend, Earl Gadd, the family he had spent so much time with as a teenager, was selling his ranch. Earl was very sick and thought he would not be living much longer. He was no longer able to do the work required to run the ranch. Pat wanted to buy the ranch, but the price of $45,000 was just way out of reason, he thought. The ranch ended up selling for $85,000 to the owner of the K Ranch, a neighboring ranch.

※ ※ ※

Grace and Lonnie were expecting a baby in April and asked Evelyn to come up and sew maternity clothes for Grace. She would like to do that, but she planned on coming back to the ranch after Tim's wedding to care for things while they were on their honeymoon. Nancy asked to be taken out to the bus stop, as she planned to spend the winter in St. George, Utah. Evelyn just had too much happening too fast in her life, and she wanted to spend some quiet time at the ranch until time to go out for the wedding.

After Tim and LaRue returned from their honeymoon, Evelyn drove her old blue pickup to Boulder. She planned on spending the winter there, as she didn't think it was a good idea to stay at the ranch in the way of the newlyweds. She took her sewing machine with her. Tim Mantle and LaRue Betke were married in a church ceremony in Boulder. There was a small reception for them at Rex and Queeda's

From left to right:
Pat Mantle,
Lonnie Mantle,
Tim Mantle, and
Rex Walker

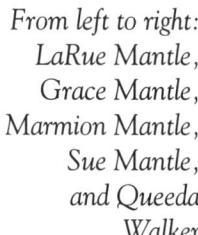
From left to right: LaRue Mantle, Grace Mantle, Marmion Mantle, Sue Mantle, and Queeda Walker

house afterward. On their honeymoon, Tim and LaRue stopped and visited Potch and Charley in Dickinson, North Dakota.

In January at the ranch, a major setback happened to Tim, who was now running the ranch. Bob Pitner, a ranch hand and neighbor, was in a bad wreck with Tim's truck. It totaled the truck, but most tragic of all, several of Tim's favorite and best ranch horses were killed in the wreck. On this ranch, horses were the key to everything done on the ranch, and it was almost impossible to replace them with good horses that could work in this rough country. Tim bought a new Chevrolet truck to replace the old one, but the horses he would just have to replace over a long period of time.

Not content to just sit around and wait for spring, Evelyn got a job. One day she was in the Singer sewing machine store and overheard a saleslady saying that the person who had been making their covered belts and buckles had quit. Upon inquiry, she found that the Singer company furnished all the materials and took all the orders, so if she took the job, she could do it at home. She would be paid by the piece, pickup and delivery at the store once a week. Singer gave her a test item to cover and found that she was fast and good and hired her on the spot.

Evelyn set up her equipment on a small table beside her sewing machine in her tiny bedroom at Queeda's house. The work rolled in. She was not able to bear to leave anything undone, so she worked long hours. She also helped with the children, the cooking, and any other jobs around the ranch. No amount of begging on Queeda's part to not work so hard had any effect. She was hard-pressed to get all the work done that Singer brought her, and she made very little money at it. However, in all her life she had never had the opportunity to earn money that was her very own, and she loved the feeling.

Evelyn found that she was going to need her birth certificate for bookkeeping purposes. She asked her cousin, Eva, to please try to get her birth certificate from the courthouse in Cicero, New York, the town where she was born. Eva had to work long and persistently, but finally got it done. Now Evelyn felt like she was more qualified to do business in her own name, and it looked like it had come to that. She was pretty sure she would not be making her living on the Mantle Ranch any more.

The good news was that Grace's mother could come to Wyoming to be with Grace for the birth of their baby. Evelyn could go right home to the ranch rather than going to Wyoming to Lonnie and Grace's house. She spoke to Singer, and they asked her to please finish all the projects she had pending and return everything, and they hoped she would be back in the fall.

Disturbing news came out of Montana. Dinosaur National Monument had sent a representative to find Charley and get him to sign permission for the ranch to be surveyed for an appraisal and possible buyout. They called Evelyn with the news that they had his permission and wanted to do the survey in January. She knew that an appraisal done in the barren month of January would be as low as it could ever be, so she refused. Then they went to Tim, and he told them they could do it in June, but no sooner. Evelyn was devastated to think that the ranch was going to be condemned and taken by the government. It seemed her life was continuing to be just one crisis after another.

※ ※ ※

Pat had just opened a new Sombrero Stable in Grand Lake, just outside the west side of Rocky Mountain National Park. His 7-11 Rodeo string had grown until he even got a contract to furnish rodeo stock at the Rangely rodeo that year. Evelyn certainly did not want to miss that. Her blue pickup was getting in bad shape. It was thirteen years old and had passed its life on the rough, muddy roads of the ranch. She didn't know what she would do when it finally quit for good. Meantime, she was glad to get back to the ranch for a while.

17

Charley, Jr., Fall 1962 to 1965

CHARLEY AND POTCH FINISHED ROUNDING UP the wild horses the summer of 1962. It had seemed an impossible task when they started out. Now ranchers could once again put cattle on their range, and there would be enough grass and water to raise and fatten cattle on. Sad at the passing of the wild horses on open range, the two men finished their job regretfully.

One day a letter came that was of great interest to them. The Cheyenne Frontier Days rodeo committee had heard that they had some wild horses available for sale. Could they possibly deliver six of them to the rodeo arena in Cheyenne? They said that they must insist that the horses be wild—the wilder the better. Potch called them and made the deal. The horses would be delivered the next day. Jack had a broken leg and wasn't any good around the ranch, so he was elected to drive the horses to Cheyenne. Everyone agreed that he should get the check before he unloaded and then take it straight to the bank to cash it.

Grinning wickedly at each other, Charley and Potch nodded in agreement. Sure enough, they had some really wild horses for them. The horses were gathered with the help of an airplane. They were gathered, loaded and trucked to Cheyenne all the same day. Potch cautioned Jack to be sure to collect the check first thing and take it right to the bank and cash it.

The event—Wild Horse Race—goes like this: A horse is crowded into a bucking chute. The only thing on the horse is a halter and lead rope. Teams consisting of three cowboys are assigned to each chute. The only advantage given the cowboys is a grip on the halter rope. At the gunshot, the chutes are opened and time begins. The winning team must control the horse enough to get a saddle on it, then one of the team has to mount and ride back toward the chute they came out of, where they must ride all the way across a white line that has been chalked across the arena. These wild-eyed, cucklebur-infested, mature wild horses threw cowboys off the halter ropes like shelling peas. That bunch of wild horses that day squealed, stomped the cowboys, kicked them, and actually chewed on them. One team finally stopped their horse long enough for a man to grab his ear and twist and bite it. The horse bit him in the leg and loosened his grip, but another man grabbed an ear just in time to keep it from escaping. Meanwhile, the man with the saddle got kicked and mauled but finally got the saddle on. It was backward. He gave his last bit of strength to turning it around, and fastened the cinch. The horse threw himself down, so another cowboy jumped in the saddle while he was down and the race was on. Pitching and bellering, the horse ran the length of the arena—the wrong way. The crowd cheered as he turned and ran toward the chute. Another team had caught their horse's rope and were earing the horse down when he got away and ran in front of the mounted bronc. The bronc spun around and followed him—once again the wrong way. The whistle blew and time was up! The wild horses continued to be the winners every day of the rodeo. That year of 1962 Lonnie rode broncs in the Cheyenne Frontier Days rodeo. He saw Jack Eaton there, on crutches, and they watched the Wild Horse Race together. Jack and Lonnie reported back to Potch: "They were little wild mean sons of bitches and ate the contestants up." Charley and Potch roared with laughter as the word reached them.

After the wild horse roundup was finished in the fall of 1962, Potch, Marmion, and the two girls moved to Halliday, North Dakota. Potch went to work on a ranch owned by his friend, Joe Chase. Potch and Joe ran some cattle together on the reservation there. Grandpa stayed there with the family for about a year, passing his time making jerky and horsehair ropes with his little assistant, Cammie. He held Chari and sang his cowboy songs and rocked her whenever he was in the house. The winter of 1963 was bitter cold. He would take the milk bucket and go out to milk the cow in her little shed about fifty yards from the house. By the time he got back to the house the warm milk had turned to ice. It was just too cold for him, and he told them he would have to go someplace warmer to survive. Joe and Potch lost about half their herd that deadly winter.

When spring was near Charley built a hothouse and started growing garden plants. He planted a huge garden. It was quite a walk from the house, making it hard for Marmion to tend it and still keep an eye on the baby. Still determined to find a good country to ranch in, Charley was preparing for his next adventure. He spent much of his time studying his Spanish dictionary. Cammie didn't like sharing her time with that book, so she hid it behind the refrigerator. Grandpa always knew where to find his book when he wanted it. He loved babies, so he was more than happy to take care of Chari. He said she was "Bompie's" little baby.

The week before Easter 1963, Queeda loaded up Cindy and Justin in the old station wagon and drove to Halliday, North Dakota, to visit Potch, Marmion, Cammie, Chari and her father. She wanted very much to see her older brother after such a long time, and she wanted to meet his wife, whom her mother had spoken so highly of. She wanted her children to know their cousins so they could be friends throughout their lives. It proved to be a wonderful visit. Marmion and Queeda became acquainted and would remain friends throughout their lives. Marmion is a wonderful cook, and they had meals at home and meals on a blanket out on the ranch. She seemed able to feed people well just any old place. The kids played together as only cousins can do. Queeda found that Potch did not like to be called by this nickname, but liked to be called Charley. Since Charley had become Grandpa, it was an easy transition to make.

Grandpa was in his own little heaven, having four cute little kids to be Bompie to. Queeda and her father had a good visit, but it was a difficult one because of how hard of hearing he had become. He told her about his adventures in Brasil and in Australia. He didn't like either place as a possible ranch location for his boys, but he was well satisfied with his adventures. Of all the things he had done, running wild horses was the one thing he liked to tell about above all else. Grandpa said he liked living near the reservation like they did. He said he liked Indians, and he especially liked to see the Indian kids play. They would take sides—one side was cowboys and the other side was Indians. They would duck and dodge and ride their stick horses fast and shoot each other with make-believe pistols, rifles, and bows and arrows. He said every one of the little boys claimed to be a descendent of the Indian who killed Custer.

Grandpa wanted to come to Colorado and see Pat and Rex's dude horse operation and visit the family, so he made plans to come to Colorado that summer.

Queeda had promised Grace they would be in Riverton with them for Easter, so regretfully they left the family in Halliday. They arrived at Lonnie and Grace's house near Hudson just in time to help color Easter eggs. Some neighborhood kids had gathered in, and the kids painted everybody and everything by the time they got the enormous number of eggs painted. They were artistic only in the eyes of the painters! Cindy and Justin were in the middle of it and happy as hogs in a wallow. The next morning the neighborhood kids descended on the house and everybody hunted eggs. The Easter Bunny had hidden them in some very unique places. However, among this group of cowboys and Indians, the Easter Bunny was easy to track in the skiff of new snow, and the eggs were all found—maybe. Kail kept stuffing eggs in his mouth until his eyes bugged out, so those couldn't be counted. Grace was expecting their baby just any minute, but she still participated in separating fights over eggs.

Cindy threw such a magnificent fit that it went down in Mantle history. The Easter Bunny hid one egg just out of Cindy's reach in a tree. She saw it, so that made it hers, and nobody else could touch it. Nobody would get it down for her. Stomping and screaming, with hot tears squirting out of her eyes as she circled like a panther under the tree, she finally laid out flat, pounding and kicking as she vented her disappointment.

Grace was famous in the area for being a good cook and never running out of food. Half the reservation showed up, it seemed. When the hunt was over, she kept setting out food until she had fed the whole crowd of people who came by.

Lonnie and Grace drove Queeda around and showed her their operation. There were little pastures with Lonnie's cattle all over that big reservation. She wondered just how a roundup could ever be done to brand or breed the cows or gather them to sell. There were no corrals, just whanged up barbed wire fences. Some cattle seemed to be just pastured in the bar ditches along the dirt roads, with no ends on their "pastures." These were referred to as the "long pastures." One night they went out to eat at the "most fabulous steak house." It was in the tiny town of Hudson and didn't look at all like a famous steak house, but actually it was. The steaks were out-of-this-world-delicious! In the years ahead, Lonnie and Grace would have their annual Christmas party for their employees there. It was a most anticipated event!

Calls from home for Queeda were getting rather urgent. They had been gone for three weeks: everybody at home was getting hungry and their clothes were dirty, and they were sick of milking the cow. Cindy and Justin would have liked to stay forever in this fun place. Uncle Lonnie was just as much fun as Grandpa had been, and you got to do anything you wanted to in Grace's house. But they did have to get on home. The trip had been great fun.

※ ※ ※

Back in Montana, Grandpa got very sick with a gallstone attack. He was in so much agony that Marmion couldn't stand it any more, and she put him in the hospital in Halliday. He was furious with her about this because "the only thing hospitals are good for is to die in." He was suffering terribly from the pain and was very feverish and sick, but as soon as he found his clothes he escaped from the hospital. He was afraid that Marmion would catch him and put him back in the hospital, so he left for Glendive, Montana, in his Jeep and took refuge with Jack Eaton.

Many of Joe and Potch's cattle died that terrible, bitter winter of 1963, then the winter of 1964 finished wiping out their entire herd. The cattle could not be reached to save them for the snow, ice, and deep crusty drifts. So Charley, Jr. moved his family back to Glendive in 1965. North Dakota had legalized gambling and he worked as a card dealer during the winter months.

18

The Mantle Ranch, 1964 to 1965

BY THE SPRING OF 1964, the Sombrero Ranches' horse herd had reached 200 head. One day in May, folks in Boulder County felt like they had returned to the Old West. A herd of two hundred loose horses driven by thirty riders and guarded by the sheriff and several highway patrol cars was herded down Highway 36 from a pasture outside Lyons to Sombrero headquarters on Cherryvale Road on the eastern edge of Boulder. It was a thirty-mile trip, and the mass of horses were strung out, filling the highway and stopping traffic while they passed. The cowboys consisted of bankers, lawyers, newsmen, doctors, and college kids. All had been invited by Pat and Rex to join the drive. The horses were being brought from winter pasture to the corrals to be shod and ridden and generally groomed up for their summer work at stables, scout camps, church camps, and rodeos. Hats were lost, scarves and coats lost, and even a few cowboys got so far behind they felt lost. There was great yelling and waving of hats to encourage the horses, sounding a lot like outlaws of old. Stan Johnson, George Nistico, and Dan Nolan had the best yells because they were experienced.

Upon arrival at the ranch, the herd of horses was corralled, watered, and hay thrown out for them. Then the cowboys unsaddled their horses and turned them in a separate corral. As the sweaty horses would roll in the soft dirt, the men would brag about how much their horse was worth, figured on how many times it rolled all the way over to the other side. A big bucket of cold water with two dippers in it showed up and everybody drank deeply. Then they all walked bow-legged up to the house, where Queeda and Evelyn and some of the wives of riders were beckoning them to come eat. A couple of big picnic tables laden with heaps of food for the hungry riders were set up in the yard. As the supply would dwindle, more food would show up, until finally everybody was full. Only then did numerous apple pies show up with a big pot of steaming coffee. Everybody laid around on the green, cool grass of the lawn at the old two-story white farmhouse and rehashed the day. Boy, was it ever a good one!

※ ※ ※

By the time spring came, Evelyn was very eager to get back to her quiet ranch home. She had had more than enough of the noise and bedlam at Queeda's. She finished up all the projects she had from Singer, gave them notice that she was leaving, and returned the buttons, belts, and equipment to Singer. They asked her if she could possibly resume her work in the fall, and she told them she would let them know.

Evelyn was sure Tim and LaRue would want to spend time together riding and working the cattle, and she could keep the home ranch going. She remembered how much she had enjoyed working with Charley before their children were born. She eagerly drove her aging pickup back to the canyon home she loved.

She arrived to find everything planted and growing, the yard watered and doing beautifully. LaRue loved to paint things, so she had cleaned up the yard and set out interesting items around it, and painted, and fixed up the house. It looked like it was going to be an easy summer. Not so! The drought was still intense, and it was a constant fight to keep the orchard and garden and yard alive. Evelyn made a trip to Summer Camp every week to take fresh vegetables and fruit to Tim and LaRue. On one of these trips in August, the rear end went out on her old pickup and LaRue had to drive her home. Evelyn left her truck with Tim to fix.

Lonnie and Grace came by, wanting Evelyn to babysit while Grace had a substitute teaching job for a couple of months, but before she could get up there, Lonnie had a mishap at a rodeo finals day in Craig. When the scores were all totaled up, he won out as all-around Western Slope cowboy, but the bad news was he threw his elbow out of joint. With his arm in a cast he couldn't do much but babysit, so Evelyn didn't feel the need to get up there before Grace's job was over.

Queeda, Cindy and Justin came to visit at the ranch in September. As usual, they had a great time and didn't want to leave. When Queeda left, Evelyn went as far as Hayden with her. There they visited with Ruth Fulton. Evelyn had boarded with Ruth when she was in high school in Hayden. Cindy sat on the hamper with her arms folded looking all around, and remarked, "Oh! It's so nice to see such a pretty, clean bathroom. Granger, you know, yours isn't very nice." "Granger" was Cindy's name for Evelyn.

It was always a lift to Evelyn when Queeda and the kids came over. Both Queeda and LaRue were expecting babies in the spring. LaRue wanted Evelyn to stay with her at the ranch that winter, so Evelyn wasn't planning on going to Boulder. She figured to try to be with Queeda in the spring when her baby was due, if she could get out then, but the weather and condition of the road would decide that.

Evelyn wrote to her cousin, Eva:

Cindy's latest: She begged me to teach her to knit. I thought it would be a passing interest but she sat out under the lilac bushes and knit all the while she was here. Then she wanted to know what you did when you wanted to stop so I showed her how to bind off. At supper she said, "Granger, when you get that last loop, what do you do with it?" I said, "Just break the yarn and pull it through; bring it here and I'll show you." "Oh," she said, "Justin took the end and just run with it and now I don't have anything left."

Time marched on, and the truck was still not fixed by mid-November. Evelyn knew they were very busy, but she needed the truck. It had been more than she could do to get fruit to the house from the orchard, carry gas to the pump, move pipe, and all the other things she needed the truck for. Then came hunting season, and Evelyn was trying to keep camp and cows intact while the kids were at Meeker at Elk Camp. She was worn out and angry.

Evelyn wrote to Eva Nov 25, 1964:

On November ninth the drought ended with a bang. Snowed 15 inches here and belt buckle-deep on the mountain. November ninth

Tim and LaRue started home with a new milk cow and two barrels of gas. A hind wheel came off on top of the mountain. They were going slow so nobody hurt, but had to borrow a truck to come on in. They hurried back, repaired the rear end of the pickup, drove 100 yards and the front end went out so that by Friday the thirteenth, when the real storm hit, they were still trying to get mobile again. Meantime my pickup is broke down at Summer Camp with the rear end out of it.

Evelyn was snowed in alone with all the deep snow and all the problems it presented, for ten days. There was not even a horse there. A hired man arrived at last with the horses and cows that had been snowbound on top, and Tim and LaRue came in their pickup. Some road workers on top just leaving with their caterpillars broke them through the eight-to twelve-foot drifts.

Evelyn wrote to Eva:

A Dinosaur Monument road man drove in. His daughter is married to the hired man so trip really to check on them but official excuse to bring the key to the road block they set up across the road over the winter. [Dinosaur Monument began at this time to close the west access road to the Mantle Ranch for the winter] *We are glad to know the road is passable because we still need grain and grocery supplies. We were supposed to be at Queeda's for a family get-together, but this branch will not make it now.*

They did make it to Queeda's for Thanksgiving, but late. The two roasted turkeys were good anyway all day Friday while the family visited. Lonnie, Grace, and Kail came, too, for the weekend. Right after Thanksgiving Rex, Queeda and Justin, with Justin's friend, Joey Walker, left for the national finals rodeo in California. The location of the finals had just been changed from Dallas, where they had attended every year. Cindy stayed with neighbors and continued school. Pat looked after the place till they got back on December seventh.

While Evelyn was in Boulder Barry Ledford brought his fiancée by to meet everybody. His sister, Cynthia also announced her engagement. Evelyn mailed to Eva a beautiful suit she had made for her while she was snowed in. She was quite pleased with how it came out, and believed Eva would like it.

Evelyn spent the rest of the winter at the ranch. She wanted to be there for LaRue when her baby was born in May. Meantime, Freda, Rex's mother, had written that she would come and take care of Queeda when her baby was due in February. It was a good thing Evelyn didn't have to get out, because the winter was blustery and cold, and it would have been very hard for her to do that. She waited anxiously for word that her daughter and the baby were all right. Tim brought in a letter from Rex and one from Freda on one of his frequent trips to town on business. Everybody was fine, and she wasn't to worry. Freda Kay Walker was born on February 12, 1965. Boulder was buried in a raging snowstorm. Grandmother Freda and Grandpa Ross took care of the children and the household while Queeda was in the hospital, and stayed for a week more to be sure everything was going well. Freda bundled up Cindy and Justin against the bitter cold and brought them to Queeda's room at the hospital to see the new baby. Humbled, but with eyes shining, they welcomed their little sister.

Tim drove LaRue out to stay with their friends, Dean and Bea Kady, when it came time for the baby to be born. LaRue endured two days of excruciating pain delivering their little boy. On May 5, 1965, Dean B. Mantle was born. In about a week, they came home with baby Dean and were very thankful that Evelyn was there to help and reassure them.

19

Mantle Ranch Dilemma, 1965

IN 1964, THE NPS BEGAN ACTIVELY trying to obtain the Mantle Ranch—by whatever means required, it seemed. Permits were cut, impossible rules for grazing the Mantle cattle were initiated, and a high-powered telescope was employed at the fire lookout station on Roundtop Mountain. The Mantles were under constant surveillance.

Charley came home from Montana to the Mantle Ranch the spring of 1965. He submitted an application for the exchange of the eighty acres of uninhabitable cliffs the National Park Service claimed he owned for the 80 acres of bottom land he had in good faith claimed under his homestead. This application disappeared, never being acted upon by the National Park Service until 1978.

On November 26, 1978, an article by Zeke Scher appeared in *Empire Magazine*, part of the *Denver Post*. Scher had been to the ranch, and had visited Evelyn in Boulder to get the details. It was titled, "The Last Ranch in Hells Canyon—After 60 years Charlie Mantle's homestead may be grabbed by U.S." It was a powerful article, describing a seemingly insurmountable problem for the Mantle family. The following is quoted, with permission from the *Denver Post*, directly from the article:

> The U.S. government, in homestead laws dating to 1862, permitted Mantle to "prove up" 180 acres, which just about covered all the flatlands he was ranching on the south side of the river at the mouth of Hells Canyon.
>
> North side of the river was 'up in the rocks' – high cliffs – and into the 1930s was marked on General Land Office maps as 'unsurveyable'

On April 11, 1929, Mantle filed his application. His address was Youghal (pronounced you-all), Colo., the closest post office (20 miles by horse trail, 25 by road). The legal description of the land was prepared by a surveyor based on the land office maps available.

Records of the U.S. Bureau of Land Management (BLM), which replaced the General Land Office in 1946, show the first survey of Mantle ranch area was in 1881, with resurveys in 1910 and 1926.

Mantle's application ran into an obstacle. Some 11,991 acres along the Yampa River had been withdrawn from entry and settlement in 1919 as power site reserves by the government. 'But in early 1931 the government received the legal opinion that Mantle could homestead the land subject to any federal rights.

Entry was allowed as of May 6, 1931, and on Jan. 26, 1932, President Herbert Hoover signed a diploma-size document for Mantle, a patent to the 160 acres 'according to the official plat of the survey of said land on file in the General Land Office.'...

National Park Service policy has been to acquire – buy out – private inholdings. Subtle—and not so subtle – pressures to sell were applied on the Mantle family – and were emphatically rejected. Other ranchers gave in....

Mantle had acquired additional land in the area for his cattle operation – on Blue Mountain and Red Rock Canyon. But in checking survey maps, Mantle discovered the homestead description in his patent had been partially wrong – for more than 30 years.

On May 25, 1965, he filed an application with the BLM to amend the patent:...' I assumed the description was correct until a survey was recently made. This differed entirely from the old corners, where they could be found.

"During the past year I was informed that the land described in my patent didn't fit the land I had located upon, built my home, cultivated and improved. The land in my original description is mostly high, solid rock cliffs.'

He proposed to convey back to the government the 80 acres of cliffs north of the river in exchange for two 40-acre sections on which the Mantles lived and worked for 40 years. Those sections were contiguous with the 80 acres correctly described in the 160-acre patent.

> No action was taken on the application. Park Service files show efforts continued by the government to acquire the property, a 1946 appraisal by a Grand Junction firm fixing a value on the ranch at $52,600....

It was 1974—nine years—before a response to Charley's request was received from the Park Service's regional office in Denver, but the land exchange wasn't fully approved for four more years, when the article by Zeke Scher, or perhaps it was just a coincidence, brought immediate action from the National Park Service. In 1978 the land exchange was finally approved. Unfortunately, Evelyn did not live to see this accomplished, nor did Charley.

After filing the application for exchange in 1965, Charley had the necessary papers drawn up conveying ownership of the Mantle Ranch to his five children. He was washing his hands of this ranch he had loved so much and worked so hard to make a living on for the last thirty-nine years. Stan Johnson drew up the papers and put them on record in Boulder. Since Tim was living on the ranch, he took on the job of continuing operations. Besides dealing with the Park Service, he was very busy doing all the other necessary work it takes to run a cattle ranch.

Evelyn had made out all the papers for the land exchange, as well as doing all the paperwork for change of ownership of the ranch. She watched passively as the transfer of the ranch took place. Silently she suffered, as not Charley, not any of the children, asked her opinion or thanked her for the gift of the ranch she loved.

※ ※ ※

Charley spent the summer in Colorado. He loved being around the Sombrero Stable in Estes Park. He was always subtly teaching the employees how to better handle the horses. One young man named Gary Burkholder was the official packer for the company. He was pretty good, but it seemed that his packs were always turning on him, requiring repacking a horse during a trip. He had studied books and listened to other people tell him how to do it, but he just couldn't seem to get it right. The diamond hitch was the most popular and most effective means of tying down a pack load on a horse. One day he noticed Charley leaning against the corral fence watching him. He asked him if he knew how to do this right. Charley said he did, and he would show

Gary how. Charley packed the panniers tightly, then hefted them to see that they were of equal weight. He hooked them over the packsaddle and piled the rest of the load on top. He threw a tarp over the load and told Gary where to tie the end of the lash rope. Then he walked Gary through all the necessary maneuvers with the rope to lash down the corners and firm up the load. He told Gary to give the lash rope that last yank and tie it off. Gary stepped back and gloated over the perfect diamond hitch he threw on that packhorse, and he told for years of the old cowboy who had taught him how to do it.

Sometimes Charley would just sit on the front porch of the stable and watch in amazement at how the American public came dressed for a horse ride, and how they rode. He refused to push the fat people up into their saddles, but he truly did like to see it done. He would chew and spit and shake his head in disbelief. He played with the five grandchildren and especially enjoyed holding baby Freda. Cindy enjoyed showing off for him by riding her pony. She taught Twister to rear up on his hind legs like Trigger, then charge at top speed up through the trees and rocks behind the stable. Justin rode too, but deliberately, with a purpose.

In the early fall Charley bought a Willys Jeep—a 1964 model, brand new. It was gray, with a cloth top. It looked just like the old yellow Jeep, except for the color. It also jumped like a grasshopper when he shoved down the gas pedal then quickly released the clutch. He fixed it all up with his camp box and bedroll and took off for Mexico. He had in his mind to go to the same places where he and Evelyn had gone in 1950.

Cindy Walker on Twister

Charley's new Jeep

Charley wrote from Mexico:

December 20, 1965
CT Mantle Conocido
Baca Choix
Sinaloa, Mexico
To: Mrs. Rex Walker
1300 Cherryvale
Boulder, CO
Hija querida. Felices Pascuas   Diciembre 20th
I thought I better write and let you know where I was. When I first came down it was so hot I couldn't take it in the low country, so I come up here in the mountains. It has been real nice but the nights are getting a little cold now. I think I will go down on the coast pretty soon. I like it up here. I have met only one man who can speak English. He cant say much so I have to talk Spanish or not talk.

I am about where the three states corner on the Rio Fuerte.

I guess Nav and Grace have got their nene by now. Shore would like to see that little smiley Freda. Hope everyone is well.

I don't know when this will get mailed. I am sending it out with a Mexican that is going to Los Mochis. If you would write me at this, Baca Choix – Sinaloa, Mexico I will get it sometime. Dad

Evelyn came to Boulder in the early fall of 1965. She had a determined, driven look about her and announced that she was going to New York. Queeda was afraid she meant to stay there, and she would never see her again. She had her birth certificate now, so she bought a plane ticket and made plans with cousin Eva to visit the family in New York. Queeda gave her a pocket knife and a handful of matches. It was a Mantle rule that you never left home without your matches and your pocket knife: you never knew when you might run into trouble, and with those two things you could survive almost anything. Evelyn spent Christmas 1965 in New York.

The Christmas festivities were held at Eva's daughter, Barbara's, house. All the huge extended family came—Cousins that Evelyn had never met, as well as people she had known and loved dearly from her childhood. Eva's husband, Billy, was getting old and was not well. His memory, as well as his body, were fading. Eva had told everybody where Evelyn lived and what a Wild West place it was and very difficult to survive in. Evelyn was a true hero to them, and they begged her for stories of her life, which she told far into the evenings.

From left to right: "Cousins"— Satie Mennon, Evelyn Mantle, and Eva Reymore

20

Lonnie's Shoeing Chute, 1965 to 1966

BACK OUT WEST, MICKIE ANN MANTLE was born in Riverton, Wyoming, on December 1, 1965. Almost from the first it was obvious that she was not well. She had allergies of every sort, and it wasn't long before they realized that she also had asthma. It was so bad that every member of the family had to change out of their outside clothes and take a shower, washing even their hair, every time they came in the house. It was a very difficult life for the whole family. Grace had her hands full with the baby and her other duties as nurse, cook, secretary, and ranch hand.

Lonnie's horseshoeing business was growing by leaps and bounds. He even got a contract for shoeing all the horses for a guest ranch out of Dubois, Wyoming. He was using his shoeing chute almost exclusively by that time. In February 1966, the Grand Teton Lodge contacted him and worked out a contract for him to shoe all their horses in the spring. Charley had returned to Montana from Mexico earlier in the spring, and by May, he had come down out of Montana to spend a little time with Lonnie's family. He went with Lonnie to Grand Teton and set up camp and cooked for him while they were there. Lonnie shod the entire herd of horses in only a couple of days with his chute, and that impressed everyone. A few years later, the lodge ended up buying their own shoeing chute from Lonnie.

During the winter of 1966, the Grand Teton Lodge got many more reservations than they expected. They didn't have enough horses to serve that many people, so they contacted Lonnie about leasing some from him. Lonnie had been around Pat and Rex's operation enough that he knew exactly what was needed, so he set out to buy enough

Lonnie Mantle on Marcell

Photograph by Will Brewster

horses and tack to fill that contract. That was the beginning of Wyoming Horses, a large horse rental outfit that is still in business.

Lonnie had quite a time getting enough horses for the lease. He spent a few months going to every horse sale in Wyoming and the surrounding states, buying every horse he thought would work. Then he needed a truck to haul them in! He borrowed a semi-trailer truck from his brother, Tim, in Colorado and headed for Nebraska to buy a used rig he had heard of for sale. The bank in Nebraska had never heard of Lonnie and he had no collateral, so in order to get a check cashed he left his rifle, tools, and whatever else the bank could use for collateral. He got the final paperwork done for a used semi-trailer truck and trailer large enough to haul horses in. The check had cleared by the time he came back through, so he retrieved his hocked items and left for Wyoming.

Lonnie took Kail and drove one semi-trailer truck, while Grace and baby Mickie manned the new semi. On the way to Wyoming late at night, Grace was pulled over by the Nebraska highway patrol for not having a proper taillight. Being a newly purchased semi, the papers of course had to be called in to be sure the outfit was not stolen. Mickie wailed pitifully the whole time. Finally Lonnie and Kail came back and helped to straighten things out. Lonnie fixed the burned-out bulb and the patrolman said to get all the other items fixed at the next town, which was in Wyoming. Boy were they relieved to cross the Wyoming state line!

Lonnie was so busy and the baby so sickly that Charley decided to stay around for a while that summer and help out. He was sure Mickie had tuberculosis and worried constantly about her. In his lifetime he had seen a lot of tuberculosis and black lung, so he was convinced she would die. He figured he could at least do chores to help them out. He and Grace went in the milk-selling business. Grace and Lonnie owned the cow and feed, and Granddad milked the cow. Grace and Granddad both strained and sometimes ran the milk through a separator. They sold jars full of milk and also made butter. They would sell their merchandise, then take the money and buy whiskey for their evening cocktails.

Charley taught Grace how to garden. They tilled and mixed in fertilizer and tenderly planted. It turned out to be a beautiful and productive garden. He also went to the river, dug up currant bushes, and planted them near the house. Grace used the fruit from them to make jelly and syrup. The garden and currant patch were the envy of the community. Grace was a world-class cook. She would bake up bread, and the family and all the neighbors would feast on hot bread and butter and jelly.

Charley liked to take all the neighborhood kids for a ride in his Jeep. Since he was a child-magnet everyplace he went, the kids flocked around him. The mothers feared for their children's lives because Charley drove where he looked, and sometimes that wasn't on the road. He would even let Kail, who was four, drive the Jeep while sitting on his lap.

21

Marmion to Colorado, 1966

WHEN THEY GOT LONNIE'S HORSE deliveries finished, Charley went back to Montana. In the summer of 1966, Marmion took the children to Colorado. Potch was away a lot, and she wanted to get a job, but there was nothing available around Glendive. Also, she wanted to get Cammie, now eight years old, into what she figured would be a better school, and closer to family. Queeda had invited her to come any time; she would always be welcome.

Marmion immediately started work in Estes Park. Rex and Queeda had purchased the franchise for a KOA campground, which they built next door to the Sombrero Stable. The manager had just left, so Marmion and the girls moved into the upstairs apartment over the grocery store and took over the huge job of managing the campground. It was grueling work, as she rented campsites, ran a little grocery store, and did all the cleanup of the entire campground. She also did all the purchasing and kept up all the bookwork for the business. Cammie was a great help to her, doing chores and taking care of her little sister, Chari, who was only four years old.

Queeda was at the stable working all the time, along with Cindy and Justin, who worked long, hard days at the stable, too. Cindy and Cammie got together a lot, and when Cammie finished her many chores helping her mother at the campground, she would come up to the stable. They rode as guides on rides, led ponies for tips, and crawled in the big grain bin to fill grain bags for the morning and evening feedings. All the children worked very hard and played even harder.

Pat was producing rodeos all over the Western Slope with his rodeo stock. He rode in the arena as pickup man for all his rodeos. He opened another Sombrero Stable in Steamboat Springs and also produced a jackpot rodeo at the rodeo arena there each weekend.

Rex was furnishing the livestock and producing Little Britches rodeos all over the Eastern Slope. He also did all the farming and produced all the hay for the Eastern Slope portion of their horse business. He took care of all the permits, taxes, and business details for Sombrero out of his little office at Cherryvale, making it by the stable a couple of days a week.

※ ※ ※

Charley was in Estes Park to visit once during the summer. He hung around the stable. He drove the stagecoach loaded with customers to the Lazy B Chuckwagon supper each evening. He enjoyed spending time with all his grandchildren. One day he tucked a big chew in his cheek, got in his Jeep, waved to anybody who happened to be looking, and drove off.

Lonnie told this story:

Dad drove that Jeep from Colorado to Wyoming to Montana. The highway patrol stopped him with a trailer, carrying a horse for Cammie, going to Boulder. Says the officer, "Mr. Mantle, you have a Colorado Drivers license, driving a Jeep with Wyoming plates pulling a trailer with Montana plates on it. You have in the trailer a horse with no brand papers you tell me is coming out of Montana across Wyoming and going to Colorado for your granddaughter!!" "Yup you got it all right officer," Dad replies. "Wellllll you have a safe trip, Mr. Mantle."

Cammie loved the small, beautiful horse. She named him "Montana". He had a perfect body and head, and he was very gentle. He was black and white, with a white tail. He was without a doubt an offspring of the famous Montana wild horse, The White Tailed Stud. Cammie rode hundreds of miles on her horse that summer.

Evelyn had spent Christmas and several months into 1966 on her trip back East. She did not really enjoy the casual, what was to her, uneventful life people back there lived. She had spent too long on the ranch, doing backbreaking work, ready always to face the frequent crisis inherent in that kind of life. The casual life back East seemed without purpose to her. Evelyn did enjoy being with her family, though, and especially the trip she took with Eva and Billy to Florida. She even got to wade in the Atlantic Ocean and hunt seashells in the sand. She commented in a letter home to Queeda:

> "We do a lot of visiting. If it ever gets too bad (boredom) I'll join the Cuban refugees or make things interesting some how. Everyone has been so good to me— it's just that I'm an outlaw at heart."

Wherever she chose to spend the rest of her life, it would not be in New York. Queeda was so glad to hear this that she cried.

By July 1, 1966, Evelyn was back at the Mantle Ranch. Some things had changed in her absence. The town of Artesia had been renamed Dinosaur, Colorado. Ray Woolley, Pat's former father-in-law, had retired from the post office in Meeker. Tim was in the midst of piping the spring at the old homestead in an underground pipeline to the campground across the creek. Best of all, he was also re-doing the pipe that formerly brought water to the house, where the gravity flow would water the yard and garden and provide all the water needed at the house. There would be no more of the back-breaking work carrying drinking water to the house. Evelyn got home in time to help fill in the trenches with rocks around the water pipe to protect it before it was buried. It felt good to work with her hands again.

The ranch was always a place of rescue for tourists. Out of gas or broke down or lost, when they needed help, people found the ranch. On July 9, thirty-five boaters came paddling and wading down the river. They had torn up their rubber boats in the shallow water. One group took Tim's pickup to Jensen for their vehicles. Evelyn took the rest of them to Echo Park for their vehicles. In addition to this frequent rescue work, Evelyn was cooking for seven or eight workers at the ranch every day. The exhaustion this caused was becoming more than she could handle.

※ ※ ※

Tim had leveled off the campground down across the creek west of the house. He had disked the field that adjoined the campground and planned to plant it in the fall. The hustle was caused by the latest news flash: a super highway was coming through the Dinosaur Monument from Lilly Park, all the way west to Utah. It was being surveyed, and when the road opened, the campground would fill up, and at last be a profitable enterprise for the ranch. Throughout her lifetime on this ranch, Evelyn had lived through many big exciting plans, just like this one. She didn't believe a word of it.

Tim dug a well at cow camp on Blue Mountain with a backhoe. In the past, a very limited supply of water had been bailed up with a bucket from about ten feet down. Now it was cleared to twenty feet deep, and there was a hearty, steady flow out of the spring and into the reservoir below where the cattle watered. Tim brought a tractor down to the ranch property and set up with his pump to irrigate the fields. Evelyn was very happy to see all these improvements to, and care of, the ranch.

When Tim left the ranch, he took one of the milk cows to Summer Camp. He dug potatoes, and gathered squash and cucumbers to take to camp also. Evelyn could see that he was tired. In addition to the work he was doing at the ranch, he was also gathering cattle, branding, and developing water in several locations on the range.

One day Barry Ledford, his wife, and his mother, Irma came to visit. Irma was Evelyn's best friend, and they were delighted to see each other. They cozied up together and Evelyn told all about her trip to New York. Irma told her all about her daughter, Cynthia, and how happy she was that Barry was happily married and doing so well with his life. Barry had gotten his master's degree in California, and he was now headed for Florida to get his PhD. His study was on the Life Cell. The draft board was after him, and everybody just hoped he could finish his education.

Evelyn was alone most of the time the rest of the summer, so she was thrilled when Marmion, Queeda, and their five kids drove in one day. Excited, but apprehensive, Queeda introduced her mother to a fluffy, gray German shepherd puppy Rex had picked out for her in Denver. They all worried about her being alone all the time. Evelyn immediately fell in love with the puppy, as did everyone who knew him, and over time he grew into a huge dog. She named him Zorro.

Marmion and Queeda had to get right back to their jobs. They asked Evelyn if she would like the four older children to stay for a while. She was thrilled, and they set a time in about a week to come back for them. A cheer went up from the kids. Evelyn was a strict disciplinarian, and they all understood that they had to have permission before they could leave her sight, and they were to play quietly so as not to wear her out. They also knew that she would let them do almost anything they wanted to do on the ranch, including playing in the river, as long as she was there to supervise.

On November 9, 1966, Tim, LaRue and baby Dean arrived at the ranch late on a rainy night. They were determined to get out to town and vote the next day, so Evelyn kept Dean, as they all knew it would be mud up to their hubcaps and a tough trip. They were glad to not have to take the baby on this hard trip.

Margaret Rinker, Katherine Rinker's daughter, had been staying with Evelyn to help her around the place. Evelyn had gotten a deal on a huge load of lumber in Craig. It cost only twenty dollars, but it took three trips with the pickup to get it home. Next Margaret and Evelyn replaced the roof on the chicken house and granary. Margaret stayed as long as she possibly could, but soon she had to go home. After she left, Evelyn put a wood floor in the granary and replaced the broken windowpanes by herself. Now Tim and LaRue could have a dry place to store grains and feed. Evelyn was horrified that used gunnysacks cost a quarter apiece. Evelyn wrote to Eva that now she had to mend burlap sacks: " at twenty-five cents apiece I can sure sew a hole."

Evelyn was very tired and a little depressed. She knew in her heart that she was just not up to this heavy work. Besides, she was noticing some tension between herself and LaRue. Evelyn decided to move off the ranch and let the young couple live their own life. She didn't know yet what she was going to do to make a living, but she had made up her mind that she could do it. She was fifty-nine years old and making a new start. Just as she had done all her life when faced with a huge problem, SHE JUST DID IT!

22

Grandpa In Montana, 1966

UGUST 22, 1966 CHARLEY WROTE to Marmion and Queeda at the stable in Estes Park:

"To All my Dear girls, and one Boy. I have gone to work for Jack [Eaton]." He pays me $200 a month. Not much but enough for a old codger like me. I don't do any hard work. Just ride. I am camped up the creek. One of the boys is with me most of the time. I have got gentle horses— I ride Squaw Man, Little Abner, and two horses of Jacks so I think I am safe. But I do worry about my girls and especially little Mickey. I don't think she will live long." [Charley was sure her sickly condition was tuberculosis, which he had seen the ravages of in his youth. He did not know or believe what Lonnie and Grace told him about asthma and allergies.]

Tim was up to see me. Boyd Walker came up with him. Boyd is all broke up and crippled. They are going to buy a big outfit on Douglas Mountain. ... I know that Douglas Mountain real well and I wouldn't give a wooden dollar for all of it—no grass or water either. There is some real good winter country down in that country that Rex is talkin' about but that is all together a different country than what Tim is talkin' about...

Well it is getting dark and I haven't got any light so will have to quit. I love you all, Dad"

Summer ended, and it was well into fall when Charley wrote again.

October 11, 1966 Glendive, Montana
Dear Queed,

I have put in for this 72 year old social security deal. I think it is only $35 a month but I think I just as well have it. They say I got to prove my age which I can't do. There was no records kept at that time, it didn't become a state [Utah] until 1896. They say get a school record the school I went to. I don't think it was a public school. I think the people just paid the teacher we had only three months school a year – no record of that. Get a statement signed by the doctor or midwife – they would be over a hundred years old so I don't think I can do that.

I am sending passport. I got an old army discharge some place but I don't know where it is, but I think it is recorded at Craig. I thought if you would go to the office and talk to somebody you might get something done. Shore wish you would try.

I am getting along fine here. I get $200 dollars a month.

Had a letter from Tim. He said his big deal on Douglas Mountain was all off. I am glad of that. *I love you, Dad*

23

Evelyn to Boulder, 1966 to 1967

JUST BEFORE CHRISTMAS 1966, EVELYN DROVE her old blue pickup to Boulder, where she was going to stay with Queeda for a while. She set up her sewing machine and made herself a wardrobe fitting to work in, then started looking for a job. It was the first time in her life she had ever applied for a job, and it was one of the most frightening things she has ever done. She feared rejection because of her age, but she faced the problem like she always had handled problems – head on. She applied for a job as seamstress and alterations lady at the Fashion Bar. It was a huge store with three large departments—mens, ladies, and juniors. She got the job and began work immediately, just as the Christmas rush was beginning. Their former seamstress had quit and left a huge pile of unfinished work to catch up on, in addition to the daily load. Evelyn even brought work home to do, with no extra pay for her time. She worked six years for

Evelyn with her grandchildren— from left to right: Mickie, Dar, and Kail Mantle

Fashion Bar as their seamstress. She struggled through these years, as her health was still not very good.

Every year Evelyn would take her vacation time in the spring around the first of June and go visit her family. She loved going to the ranch, and she would go to visit her Wyoming family as well. She visited Sue and Steve, Pat's son, in Meeker on every trip to the Western Slope. All the grandchildren loved her, and all of them but Steve called her "Granger". To Steve she was his beloved "Grandma." Granger was the name Cindy gave her when she first began learning to talk.

Evelyn wrote to Eva from 1262 60th St. (Cherryvale), Boulder, January 22, 1967:

> "It is a fever pitch around here. Marmie and I leave at 8:15 every morning and get home about a quarter of six. Getting supper out of the way and clothes ready for the next day and lunches in the morning keep us pretty much out of trouble. Sundays are so full I'd almost rather work.

Marmie and Evelyn were looking for a place to live. Evelyn figured they would probably rent for now, but she wanted to own a place as soon as they could find something to buy, so they would have equity in it. While looking for a home, they rented a trailer house on Darrel Lemon's farm on Baseline Road.

Evelyn found herself handling alterations for all three departments at Fashion Bar by herself. Her helper was let go. Fashion Bar began cutting back when two big discount stores went in. Evelyn's lifetime of making new clothes out of old ones was serving her well. She could alter clothes like no one they had ever seen. She was overwhelmed by the huge amount of work she had at the store, and in addition, customers and friends were constantly begging her to do private work for them. She just couldn't do it. She even had to take some of the store work home with her nights to get it all done. She did all this in her tiny bedroom at Queeda's. After moving to the trailer house she had a bit more room.

Evelyn wrote:

> I've had to learn to demand my rights and respect from the "bosses". We had it out one day. I was ready to throw their yardstick in their faces and walk out, but they apologized and one even brought me a new steam iron next day, which I had been asking for ever since

I began work. So now I feel I know the ropes and the clerks say please and will you rather than the demanding tones they were using at first—so it is more fun too. This underdog bit always did irk me and I refuse to be treated as such. Fashion Bar is beginning to sell "tent" dresses. As seamstress I do lots of adjustments, for people who cannot totally accept the tent look.

Evelyn was showing the ravages of too much work and the strain of working too long hours. She never talked with anyone of where she came from or what she had done in her life. She was afraid that somehow it might affect their perception of her work and she would be fired. At the same time she would tell Queeda, "I would love to tell those snooty women what I have done in my life and just see the look on their faces. Most of them have worried all their lives only about the impression they make on other people."

※ ※ ※

February 12, 1967
Foss Place- Montana
Dear Dotter,
Jack was down today and brought your letter. Oh I am all broke up about Mickey. [Her endless fight to live through her asthma and allergies, which Charley falsely believed to be tuberculosis.] *Don't know why it hit me so hard, I've been expecting to hear of her dying since I was there last summer. I never was so broken hearted in my life— the little thing was just skin and bones.*

Well I been down here since first of the year. I been alone all the time but I don't mind. I got a good camp, good warm house lots of hay and oats and a couple of real good horses and they are perfectly gentle. I don't have a lot to do, just keep the water holes open and pack salt. They sure eat lots of salt. This is strictly grass country – no salt feed of any kind. That is about the only trouble me and Jack have. I have to eat on him all the time, he come from down in Wyoming from a desert country which is mostly salt feed. He don't seem to know the difference but he is beginning to see the light. Jack brings in plenty of salt for the cattle now.

I ride one part of the country one day and another part the next and so on – try to see all the cattle at least once a week. If anything needs feed I bring them in. Only brought in two so far.

I didn't intend to stay here this winter but for one thing Jack was up against it and said he couldn't get any help and another thing I thought if I stayed I could help Marmie and the girls a little. Now I am glad I stayed – it hasn't been a bad winter and it is about over but I don't think I will do it again.

Well this is another day – 13th and a big blizzard on. Can't hardly see the corals about 50 yards. I run out feed the horses and hurried back. I'm not going out today. I got lots of wood chopped up so I am OK. I will just lay around today and talk to my cat.

Queeda, can you find me a little cook book. I don't want one of them big old things. We used to have some little ones at home. I think they was put out by baking powder companies. They had lots of old time recipes to make pie, doughnuts, pudding – stuff like that. I know how to make bread pudding, but biscuits don't work so good. Days like this I have lots of time to cook if I know how. I get hungry for something sweet. I got a friend at Miles City—she brought me a dozen jars of home canned fruit couple weeks ago. I shore have enjoyed it—I don't like store stuff. It don't taste right, not to me.

You ask about sending me socks and boots— I got lots of clothes—don't need anything. I bought a pair of boots about three months ago. Today is the first time I've had them on—been wearin them round today. Lots of deer and antelope round here but I haven't killed anything yet— got lots good beef.

Guess I better rap this thing up. Would like to see you all. Let me know about little Mickie—I am holding my breath. I love you, Dad

※ ※ ※

Evelyn was relieved that Grace's mother, Nellie, would be with her when her baby was born in May up in Riverton. Nellie had allergies and asthma problems and knew better than most doctors how to take care of little Mickie in the very best way. Mickie continued to be very ill and frail and Nellie would not forget to keep everything filtered and pristine for her. On May 20, 1967, Darlene Kay Mantle came howling into the world. This beautiful baby girl would keep Lonnie and Grace in unending grief and joy, but never boredom, for the rest of their lives.

Grandpa Charley wrote:

July 30, 1967
Glendive, Montana
Sent to Mrs. Rex Walker
% Sombrero Stables
Estes Park, Colo.
Dear Dotter,
Well as the old song says, I am back in the saddle again. Workin like hell and it is awful hot. I don't feel too good. I don't sleep too well. I worry all the time about little Mickie..

I didn't get my welfare check for June— don't know if somebody stole it before it got to you or if somebody up here got it. Maybe you can find out. The July check is due now. Maybe you better register it. Don't know what I am going to do yet—have been offered a good job in Canada takin care of a bunch of race horse mares. Will tell you more about it when I get more details. Dad

All at once, on Memorial Day, a great silence settled over the Cherryvale ranch. Queeda and her family and Marmion and her family moved to Estes Park for the summer. Evelyn felt lonely, but threw herself into her work. She visited the families on weekends in Estes, but mostly Marmion was too busy at KOA to spend much time with her, and Queeda was at the stable, busy all the time. The children were all over the place, playing, riding, and even taking out rides and leading ponies around for small children for tips.

When business in Estes was over for the summer, everybody moved back to Boulder and the kids started school. Marmion put the girls in school, then got a job. Cammie and Cindy went to the same school, and they and Justin caught the bus every morning out in front of the house. They loved having each other. After school they rode horses, Justin secretly rode the pigs, and they all did chores. When Evelyn got back from church on Sunday, January 22, she found that Rex had just gotten back from a trip to Craig. With him had come trucks loaded with horses, and the place was bustling with noise and activity. Marmion, the two girls, and Evelyn were living in the old farmhouse, and most of the time Pat was, too, so it was a very full house. Queeda loved having her family around her. However, Freda Kay was beginning to walk, and she would get out of sight quickly. The steep stairs were a constant attraction, so a baby gate was installed. She would slip outside every chance she got, and the family worried constantly about her getting run over or backed over in the busy driveway. Queeda was putting up a howl to move.

Pastures were rented all over Boulder County for Sombrero livestock. Rocky Flats pastureland, out west of Boulder, held Brahma bulls, horses of every shape and color, shaggy Scottish longhorn cattle, a couple burros, and all colors of cows and scrawny roping steers. Smaller pastures, and many old farmhouses with a few acres attached, were leased by Sombrero.

Short on money always, Rex and Pat decided to "give" people horses for the winter. The deal was, if you wanted a bulletproof gentle horse for your kids, you went to Sombrero and picked out a horse. They furnished you with tack and hauled the horse to your house. You signed a contract to care for the horse properly, and he was yours for the winter. Some families took horses year after year.

Rex set up a stable at Cherryvale and started renting out horse rides. Pat put in an arena, and there was a jackpot rodeo every weekend. The noise and activity riled up a good part of the neighbors. The rest came and either worked or just enjoyed the show. Queeda cooked and answered the phone and hoped to have some friends left. It was a big relief when everybody and most of the horses left in the spring to go to work in the mountains at the stables, or traveled around the state to put on rodeos.

Sombrero even set up another small stable on the northeast side of Boulder off Jay Road and 47th Street. It was an old farmhouse with lots of outbuildings, and the surrounding unoccupied small farms and creek banks were good areas to ride on. A hippie kid came to live there to help with the horses. He had a big old ugly dog that looked dangerous, and Rex told him if he was going to keep the dog there, he would need to chain it up. One day Rex drove in and the kid was hopping mad. He said, "Somebody ripped off my dog." Rex was baffled and said, "How could they rip the chain loose? It was wrapped around a big old tree trunk, then nailed down." Answer was, "No, no, they ripped off the dog, you know like stole him."

※ ※ ※

Rex and Queeda were talking of building a new house. They had a nice-size acreage about purchased over in north Boulder on Jay Road. It was a quiet, rural area, and they would probably sell the Cherryvale place for development and then move the buildings to the new place for housing for the employees, the Sombrero office, etc.

By 1967 Sombrero Ranches had grown to the point where Pat and Rex had to get more winter pasture. Pat knew of a big ranch for sale near Craig, so they went to look at it. Since Pat was well known and trusted in the area, they were able to buy it with very little collateral. The ranch was located north of Highway 40 at Big Gulch, thirteen miles west of Craig. It was 7,000 acres of rolling sagebrush country with lots of grass. Quite a bit of BLM land was attached to it, too. It would probably get too much snow to be good winter country, but it would make a wonderful fall and spring pasture. The horses could come right out of the stables into that good pasture country for rest and revitalization, after a hard summer carrying around tourists and bucking off rodeo cowboys.

Renting out hunting horses had become a big business for Sombrero. Now they could rent horses out of the new ranch for Colorado's booming big game hunting season on the Western Slope in the fall. Also, the business was expanding rapidly in summer rentals to camps, many of them on the Western Slope. Pat built some beautiful corrals, and they fixed up the old milk barn on the place for living quarters. Now in the spring they would be able to gather, sort, and get all the horses shod and ridden once before they were sent out. This one first ride always had to be made on each horse. The gentle dude horses

Pat Mantle resting at the Big Gulch Ranch

regained the spark of youth out in those wild pastures over the winter, and some of them bucked a little. One ride in the spring and a handful of oats, and they were ready to get back to their other life. They were trucked out to where they would spend the summer working. They always whinnied and nickered when they saw their stable, because they would get oats and loving care all summer.

Queeda was sad that Pat would have to move to the Western Slope. She would miss him terribly, but Pat was the only expert on the care of horses on these huge pastures, where water was scarce and had to be handled carefully. Also, if a hard winter hit, he was the only one experienced enough to get the horses through it. He would have to move to the area. He was excited to do that and was soon settled in. His rodeo customers knew where to contact him, and any other business would go through the Sombrero office in Boulder to reach him.

24

Browns Park, Jay Road, and Charley, 1967 to 1969

PRETTY SOON REX AND PAT WERE DICKERING for another ranch on the Western Slope in Browns Park, about fifty miles west of their Big Gulch ranch. Charley knew they were needing winter pasture for the horse herd, and he had told them of a location in Browns Park that would be perfect. He took them to look it over and explained how to use it to best advantage. It was not much good for sheep or cattle, but it would make a wonderful pasture for horses if used right. After Charley schooled Pat and Rex on how to use the new ranch to best advantage for horses, he drove off to Montana to spend the winter with Jack Eaton.

The deal for the ranch was held up by the reluctance of the BLM to change the permit on the public land attached to the private land from cattle to horses. The Brown Park Ranch, including Boon Draw and Thompson Basin, contained 15,000 deeded acres. The BLM ground attached to it came to about 40,000 acres in all for the horses to winter on. An agreement was finally reached, and the Browns Park Ranch came into Sombrero's possession in 1968.

Tim wrote that it was so cold at the Mantle Ranch that there was a foot of green ice covering the river. He could drive the old yellow Jeep all the way down the river on the ice to Outlaw Park to check on weaner calves there. At least he didn't have to worry about them coming out of there on such slick ice. The bad news was that usually there was so little ice on the river that it was easy for the cattle to get a drink all winter. That was a problem now in all the pastures along the river, and would take a lot of work to get the ice sanded. [So the cattle could walk out on the ice to the edge of the water to drink.]

※ ※ ※

On February 20, Queeda wrote to Charley to reassure him about Mickie, who he worried about all the time. Grace and the kids had come down to her house from Wyoming, and Mickie was growing normally, alert and happy. Mickie was getting tests and treatment for her allergies from a specialist in Denver, and the treatments had made a big improvement in her plant allergies. She was too young yet to get the series for animals, so everybody was trying to keep the dogs and cats out of the house. Hardest of all, though was keeping Mickie out of the barn and chicken house.

※ ※ ※

In early 1968, Rex sold the Cherryvale property and took the money to buy the land on the east side of Estes Park where the stable was located. They had been renting the trails on the ranch from a fine old lady, Muriel MacGregor, who passed away. They wanted to secure the Sombrero riding area forever, so they were elated to get this done. At the same time, Rex bought a school section on the back side of the MacGregor ranch. They felt that they now had a location that would serve the family well for many generations.

That was one of the most fun summers ever at the Sombrero Stable in Estes Park. Eight kids all about the same age did the work of adults and lived the life of Wild West kids. They started work at 6:45 in the morning and crashed into bed after ten at night. Anybody that was forced to sleep in howled in protest. The kids all wanted to be where the action was, down at Sombrero Stable.

Keith Hagler and his wife, Eileen, managed the stable that summer. They had three children—Mike, Cindy, and Tommy. Rex's nephew, ten-year-old Walker Weathers, from Tyler, Texas, came to spend the summer. Cammie, Chari, Cindy, and Justin finished out the crew. Queeda manned the office and cooked for the breakfast rides and steak fry rides out on the range. Customers loved this crew of competent, happy children.

Rex and Queeda, with Rex's father, Ross's, help, bought the land on Jay Road, and they started getting serious about house plans. They would have to move in the fall. Queeda's old friend, Joe Robinson, who she had worked in his restaurant for, in college, was now a builder. They signed a contract with him to build their house. It was wonderful to have a builder they had so much confidence in, so that they could be gone working at the stables and rodeos all summer and not worry.

Queeda turned up pregnant, and it was a race between what would happen first: whether the house would be finished or the baby would come. The kids would start school the day after Labor Day, then they would have to move out of the Cherryvale house, as Keith and Eileen were scheduled to move in. Of course, the Walker family must clean up the old house after they moved out of it. All this work had to be finished by the first of October. That was the peak of the mad rush of trucking hunting horses around and setting up hunting camps. As it turned out, Queeda and the kids moved down from Estes Park, moved into the new house with a limited amount of furnishings, and two weeks later the baby was born. Cody Rex Walker rushed things a little and was born on September 28, 1968. Eileen Hagler drove Queeda to the hospital, and Rex arrived in time for the birth. Evelyn and Marmion, of course, helped out through all the madness. Cody was a good baby, unaware of all the turmoil around him. Rex left almost immediately after the birth to tend his hunting camp at Beaver Creek, in the mountains near Meeker. Eileen took over the job of loading out the hunters who came by Cherryvale with their horse trailers to pick up their hunting horses.

The house turned out to be beautiful. Joe and his wife, Josephine, did the tedious detailing work by hand. There was a barn and a few corrals on the place, so the kids' horses and the milk cow were brought in right away. Everybody felt right at home after that.

Evelyn and Marmion were living in a trailer house on Baseline Road on the farm of Darrel Lemons. They enjoyed being away from the constant hubbub of Queeda's house. Potch came to visit his family for a few days, then had to rush back to his job. He was working in construction.

As luck would have it, Evelyn found a beautiful piece of land of almost two acres on Jay Road very near Queeda's house. [Joe Robinson built her house right after he finished Queeda's, and she and Marmie and the two girls moved into it in 1969.] Pat wanted to pay for the acreage and house, but his mother wouldn't hear of it. She

Potch Mantle with his family— from left to right: Chari, Cammie, and Marmion

finally let him go on the note with her so she could buy the land and build her home.

Evelyn had a big laugh when she saw that the loan was for twenty-five years! That would make her eighty-eight when she paid it off. She worried endlessly about this loan and ended up paying off the whole thing in five years. She worked at her job all day, then did private sewing after work and on weekends. She kept a little tablet of her expenses daily, even down to the price of a package of gum. There were absolutely no luxuries in her life. Her old pickup was falling apart, and many evenings you would see her out front with the hood up, working on it.

Since Chari was not in school yet, she got to go in the truck with Uncle Rex quite often. Once she went to Craig with him for three days. When she got home she was so excited telling the stories that she stuttered. Rex would always take the kids to local horse sales with him. Cammie, Cindy, and Justin often rode the gentle old horses the men were selling in the auction ring. It gave buyers confidence that it was a gentle horse, and the horses brought a little more. The kids were referred to as "sale barn rats," as they ran around through the seats and rafters, barns and corrals, and ate all the time.

It was a really tough winter. Snow and ice covered the land. Earl Vaughn, a Sombrero employee in Boulder, went over to help Pat in Browns Park. Mike Harding, another top hand, was helping Tim. Keith and Eileen Hagler had managed the stable in Estes Park that summer, and when the stable closed they moved to Boulder to take care of the livestock.

In the early spring of 1968, Charley returned to the Mantle Ranch. He had given it a lot of thought, and he needed a home. Tim and LaRue were at the ranch house, and he didn't want to stay there anyway. He had always dreamed of a home at Red Rock. It had good farm land already cleared and leveled, easy to water, best springs in the country, and provided a good view from an area protected by cliffs and hills, backed up against Blue Mountain. Tom Blevins' old homestead cabin was falling in, so when spring broke Charley moved into an old trailer house that was on the place. Water was already piped to it, and there were good, big shade trees all around it.

In March, Queeda wrote Charley that his sister, Aunt Nancy, was staying with them in Boulder and was really crippled up. She said Rex had spent all February in Roosevelt, Utah, tending an oil well that was being drilled. The bad news was that it was dry. This was his second dry hole he had an interest in out there.

Charley Mantle, Queeda Walker, and Chari Mantle in front - 1968

In April, at spring break for the kids, Queeda, Marmion and the five kids came to visit. Charley proudly showed them around his new home. He asked Queeda to check out an old schoolhouse he had spotted. It looked like it was movable and would make him a nice house. On her way home, she looked at the schoolhouse and reported it was in good condition. It was already up on a foundation, and would be easy to move. It measured eighteen by thirty feet, and it looked like it could be easily cut in half, so it could be moved on the narrow road a half at a time. Charley was sure Pat could find somebody to move it for him.

In May, ten-year-olds Cammie and Cindy got their horses and saddles all polished up and competed with some other girls for Little Britches Princess of the Little Britches rodeo, to be held at the Pow Wow Grounds in Boulder in June. Cammie, on her little black-and-white-pinto horse, Montana, that Grandpa had given her, won with no problem. Grandpa sent her a pair of beautiful braided rawhide reins he had made for her. She was so proud of them and they were so precious to her that she could hardly make herself use them.

Pat opened a Sombrero Stable in Steamboat Springs in 1968. It was located in a scenic old barn at the foot of Mount Werner. He also

Cammie Mantle Little Britches Rodeo Princess

put on a jackpot rodeo on weekends at the Steamboat rodeo arena. He became the much-celebrated resident (real) cowboy for Steamboat. His picture appeared on posters advertising Steamboat as a Western ski area. Movie stars came to meet him, and he was celebrated as being one of the last "real" cowboys in the West.

As the summer of 1968 wore on, Charley worked so hard setting up his home-to-be on Red Rock Ranch that his seventy-five-year-old body ached. He thought more and more of that big hot spring in El Fuerte, Mexico, that he had soaked his aching, tired body in. He had done about all he could here, because winter would be upon him soon.

That fall, Rex met a Mexican man and his wife who came to hunt at his Beaver Creek hunting camp. His name was Felippe Jurado and his wife's name was Carmella. They got to visiting and made a plan. Each one of the Jurado's three children and each one of the older Walker children, when they reached grade six in school, would spend the school year in the other country as guests of the family. Completely immersed in the foreign language, they believed the children would learn the new language quickly and permanently.

Felippe and Carmella visited the Walkers in Boulder. They were wonderful people, and they all became good and trusted friends. The mothers had dire misgivings about giving up their children, but they were overruled.

In September 1969, Rosa Jurado came to stay with the Walkers. She and Cindy and Cammie got along famously, and she was a delight to have. She learned English quickly and was a star student at Douglas Elementary school. She was popular everywhere she went, whether on the school bus, in class, or around the ranch. She participated in everything! The three pretty young girls were always together, a brunette, a blond, and a light brown.

Charley's hearing was really bad by now. He had gotten a hearing aid, but the damn thing went into a loud, shrill whistle every once in awhile that filled his whole head with pain. When he was out by himself, he could turn it on and hear the sounds of nature

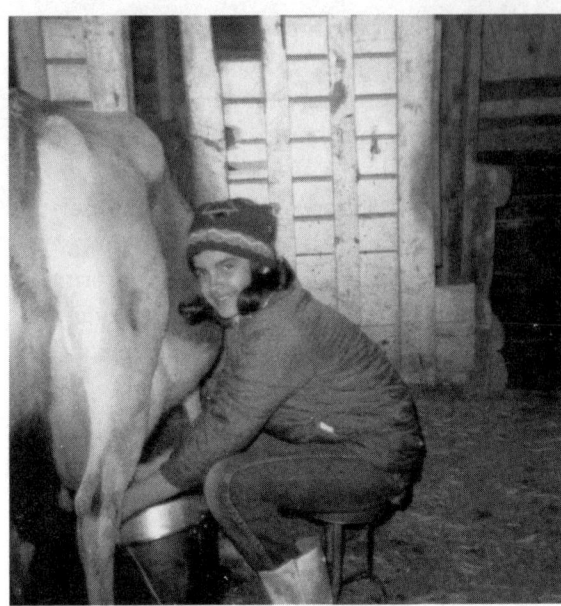

Rosa Jurado demonstrating yet another skill

he loved—water running, birds singing, quaking Aspens, wind in the trees, horses running. He could even somewhat carry on a conversation with people with regular voices, but shrill kids' voices sent the thing into its scream. He hated to go back into Mexico, not able to hear anything or talk to anybody.

One day he was at the ranch house and Tim, LaRue and Dean drove in. A young woman, who was a friend of theirs, was with them. She and Charley got to talking about how he wanted to drive his Jeep to Mexico and camp out for a week or two, soak in the mineral water, and do some sightseeing. They talked for awhile and they struck up a deal. She would like to go with him and be his ears and see Mexico. It sounded like fun.

They crossed into Mexico at Nogales on December 27, 1968. They had Charley's bedroll and camping supplies, her sleeping bag, her toy poodle dog, her .22 caliber pistol, and Charley's .22 rifle and his pistol. These guns were carefully hidden and the inspection was minimal.

The first night, the 28th, they camped on the coast around Hermosillo. On the 29th they drove almost to Los Mochis, and the 30th they stopped in Topolobampo. Charley waited there for mail from Queeda with a check in it. It didn't come, and they went to the little town of Charay for a New Year's party with some people they had met.

The next morning, Charley's traveling companion wanted to spend some time with an acquaintance she had made at the New Year's Eve party. She said she was tired of camping out, so they split up. Charley drove about fifteen kilometers to El Fuerte and camped near there. There were many Mexicans camped in the area, because there was a government giveaway of land going on around a newly-built reservoir. Charley had no idea what was going on.

Evidence showed that during the night of January fourth, Charley was sleeping in his bedroll by his Jeep when he heard someone pawing around in the Jeep. He rolled quietly out of bed, grabbed his pistol from under his pillow, and crept barefoot to the side back of the Jeep. Apparently at this point he was attacked from behind, hit in the back of the head, and killed with his own shovel.

Early in the morning on Sunday, January fifth, his body was discovered by other campers in the area. A call to the Los Mochis police reporting the crime came from San Blas, a very small town near El Fuerte. It was a woman who called it in. The police from Los Mochis came and inspected the scene and called the family.

Meantime, Charley's traveling companion had come into Los Mochis with the Mexican boy she was staying with, bringing her poodle. The dog had been shot. Both people were covered in blood, so the police threw both of them in jail for the murder of the American. They called it a crime of passion, thinking she and her friend had come in the night and killed Charley, and it was his blood on the dog. Her story was that they were fifteen kilometers from the scene of the crime, and her Mexican friend had taken her pistol and shot her dog running through the tall grass, thinking it was a coyote. They had brought the dog to Los Mochis looking for a veterinarian.

When the call came in about Charley's death, Rex took the first plane he could get to Los Mochis. Queeda called her brothers, and her worst job was to tell Evelyn the sad news.

Rex arrived in Los Mochis on Monday. He identified Charley and asked that he be embalmed so his body could be returned to the United States. Tuesday he investigated the crime scene with the police and went to see the woman in jail. Later that day, she and her Mexican companion were transferred to the 400-year-old jail in El Fuerte. The plan was that if she proved innocent, she would be released and sent out of the country when Charley's possessions were released. Wednesday, Rex interviewed everybody involved. Thursday he went back to El Fuerte, where he talked to all the campers and found that nobody had seen or heard anything, and the only person they saw around the

Jeep was the "old man." Rex got possession of the Jeep. On Friday he got possession of Charley's effects after the police made a call to Pat to get permission.

On Monday, January 13, Rex got the young woman out of jail, put her and her dog on the bus, and he drove the Jeep to Tucson, Arizona. There he parked the Jeep and contents under the care of Galloway Motors in Tucson, Arizona, where one of the Mantle brothers would come pick it up. Rex flew home.

The family wired $200 to Carruns Mortuary in Nogales to fly Charley's body to Denver. Lonnie met the plane and carried the casket to Craig in a station wagon. Tim went to Tucson to pick up the Jeep, and Pat got together a crew and went to dig a grave for Charley. Since he had so loved the Red Rock Ranch, they all agreed he should be buried there. Pat searched out a place on a little hill overlooking the valley with the yellow cliffs of the canyonlands shining in the background. It soon became evident that it was bedrock and also frozen deep, but Potch had come prepared: He got some dynamite out of his truck and set off several charges to break up the solid stone floor. After that it was easier digging, and they got a nice, deep grave dug.

A funeral was held in Craig for Charley. There was standing room only in the big hall. Several old friends, including Joe Haslem, got up and spoke of their memories of Charley Mantle. The old cowboy had not passed quietly through this land. He would be missed, but his legend lives on.

Billy and Francis Chambers, old friends of the family, gave over their ranch home to a huge reception after the services. There were well over 300 people there, and old friends got to visit quietly. Others not so quietly enjoyed the afternoon. Nobody seemed to notice that it was snowing and getting on toward night, and they had a hundred-mile trip yet to make to the gravesite. The snow just kept coming, and the dirt roads were almost impassable. Since nothing was impossible to Charley Mantle in his lifetime, certainly nothing was going to stop his funeral procession. His friends and family got him to the gravesite and sadly buried him. The tragedy and mystery of his death still haunts many people.

Those folks attending the burial were: Evelyn, Potch, Marmion, Pat, Lonnie, Tim, LaRue, Ellis Wilson, Ray Green, Sam and Darryl Steele, Sam MacIntyre, Dick Toole, Boyd Walker, Jerry Crook and wife, Monte Sheridan, Guy Urie and wife, Mike Hume, EL Buzby, Doyle Eaton, Doc Brown, Elmer Mack, Roy Grounds, Bill Chambers, Jack and Marjorie Eaton, Jim and Cathy Wright, Mike Harding, and Boots Chevington. Please excuse any omissions.

25

Charley, Jr.—Potch, 1970 to 1971

STILL RAW FROM CHARLEY'S DEATH, more bad news arrived for Evelyn and Marmion. In November 1970, Jack Eaton called to tell Marmion that Potch had been killed in a car wreck. Plans were made between Lonnie and Marmion: she would like for Charley to be buried at the Mantle Ranch alongside Grandpa Mantle.

Pat rounded up a crew and the tools they would need. They set out for Red Rock on the Mantle Ranch to dig a grave. Jim Scott, a long-time friend, reported later that it was solid rock they had to dig through. They didn't dare use dynamite, as it most surely would destroy Charley, Sr's, grave right there alongside. Commenting on the difficulty they encountered, Jim said, "I ain't never gonna help Pat Mantle dig a hole again!"

Lonnie drove to Billings, Montana, to meet Jack Eaton, and they went together to Hamilton, Montana, to claim the body. They made all the arrangements and hired the funeral home there in Hamilton to transport the casket to Zoebels Mortuary in Craig, where a funeral service was held.

It is always a sweet and amazing thing how Western people get the word around about one of their own. People showed up from all around the country. They had known Potch when he was a youngster, or when he rodeoed, or when he ranched, or when they had worked with him various places. A lunch was provided by somebody at their house in Craig after the service. Everyone gathered around, swapping stories longer than they should have. Potch's body was transported in the pitch dark of the night by pickup to the gravesite 100 miles west of Craig to the Mantle Ranch. He was buried beside his father on the Red Rock Ranch.

Evelyn buried her eldest son, who was only forty-two. It had been only ten months since she buried her husband of forty-three years. Her body carriage showed the grief she was holding inside, and her eyes showed all the agony of her soul.

After her husband's death, Marmion felt a deep yearning to be with her family, who lived in Washington State. When the girls got out of school in the spring they went to the Mantle Ranch to spend the summer with Uncle Tim and Aunt LaRue. Work was hard there, but how they loved it! They could ride horses, swim in the river, eat fruit from the orchard, and generally run wild as the wind. Tim and LaRue and Dean always loved having them, and in the years ahead they spent many happy summers there.

※ ※ ※

Evelyn paid Marmie for her equity in their house. She knew Marmion would be needing the money. Marmion wound up her employment, picked up her girls at the ranch, and headed off to Wyoming. They stayed at Lonnie's ranch for awhile, but there was no work to be had around there, so on August 1, 1970, she drove on to Tacoma, Washington. She had heard from her family that good jobs were plentiful there.

Marmion called Evelyn to report they had arrived, they had a good trip, and the girls were registered for school. There were lots of job opportunities, but the continual smog and haze concerned her. She wondered if she should have stayed in Montana where there was still some blue sky and fresh air.

By October 9, 1970, Marmie was living in a trailer house next to her mother. She was not finding work that she could do that paid enough for her to survive on, so she and the girls moved back to the West they loved. Lonnie and Grace helped her when she needed it, and Cammie and Chari enjoyed the ranch life they had missed so much. The girls spent summers on Lonnie's ranch and on the Mantle Ranch with Tim and LaRue. It was nice having a big family.

Marmion worked long, hard hours at a job, then came home and sewed clothes for her girls and always had a garden. As soon as she could, she bought a home of their own in Lander, Wyoming. She worked at the Wyoming State Training School there as the Food

Service Manager until she retired. Marmion still owns her home in Lander, and keeps in close contact with her two daughters, her grandchildren, and great-grandson.

Years after Charley, Jr's death, Marmion married a very special man from Australia named Kevin Hocking. He was a person who always brought cheer where he was, and the extended family loved him dearly. They were married at Lonnie and Grace's ranch beneath the flags of their respective countries. Kevin loved Australia, and he brought a little bit of it to wherever he was. He painted and planted a big road sign for Lonnie. On the sign was a kangaroo pointing to Lonnie's ranch. Marmion and Kevin had a happy life together until his death in 1993.

Charley and Marmie's oldest daughter, Cammie, married Jim May, her high school sweetheart. They live in Casper, Wyoming, and Jim is a welder by trade. Cammie taught herself computers and bookkeeping and used both those skills as secretary of the Mantle Ranch Corporation through its bitter battle with the government for possession and use of their land.

Marmion's youngest daughter, Chari, is an elementary schoolteacher in Gordon, Nebraska. Each summer she takes an adventure trip someplace exciting. She has an abiding love for the Mantle Ranch, where she spent so much of her childhood.

26

Lonnie to Pavillion, 1973

THE WINTERS WERE HARSH on the river between Riverton and Hudson, Wyoming. All winter a cold blue haze and usually fog laid over the valley. One year even the whiskey bottle under the seat in Lonnie's truck froze. It was not a good place to winter livestock. The family had outgrown the little house, and Lonnie was also looking around for a better place to keep his livestock.

In 1973, Lonnie and Grace found a 160-acre place with good water and lots of irrigation rights. It was out around Pavillion, just west and a little north of Riverton. Lonnie knew the area because he wintered horses out there. It wasn't far away, but it was like a whole new world in winter. It was warmer, the snow didn't lay, and the feed was strong. The fences were bad, but weren't they always? There were plenty of pastures and farms to rent for winter grazing.

The ranch house was just three miles north of Pavillion. It had been moved in by the government from the Japanese concentration camp at Powell, Wyoming, after World War II. It had been an officer's house and very nice in its time, but was now in bad shape. The ranch was a project of government. They set up 160 acres of land and a house for veterans after World War II. The project was a failure, as of course no one could make a living on 160 acres.

The house needed a lot of work done on it before they could move in, so that spring Grace's folks brought up their motor home and parked it by the new house. Grace and the kids moved into it. Lonnie spent what time he could there, but he still had to spend most of his time at the house in Hudson where the phone was. He got calls about his dude horses and from customers wanting horses shod. He couldn't afford to lose any of this income.

Under Grace's supervision, a couple of part-time carpenters gutted the inside of the house. Money was scarce, so the carpenters salvaged what they could from the interior of the house and used it in the remodeling. Grace found some used windows from the lumberyard and bought paint from farm sales. By the time school started in the fall, the house was livable and the family moved in. Each of the girls had her own room, and Kail had the whole basement to himself. Lonnie and Grace had a huge bedroom that would hold a king-size bed, and had windows looking out over their valley. It was like a castle compared to the house they had outgrown at Hudson, and "it was all theirs!"

They cleared, fenced, tilled, and planted grass seed in a generous yard area. It could be irrigated from the irrigation ditch that ran through just above it. Grace portioned off an area for a garden. Lonnie leaned on his shovel and spoke slowly and clearly, "I will get this damn garden ready, but I will never plant it, weed it, or water it. I will never put up hay, either. I've had enough of that to last me a lifetime."

When Evelyn made her yearly visit that spring she was thrilled for Grace that she at last had her very own house. Evelyn purchased a dishwasher for the new house. Grace protested, and Evelyn used the excuse that all she ever did when she visited was do dishes. It was true, Grace always had a house full of people to feed, and she enjoyed the dishwasher, with fond memories of Evelyn every time she used it. Lonnie knew what bad shape Evelyn's old truck was in, and bought her a brand new car. By her specifications it was a four-door Chevrolet with a standard shift, so she could manage the ice and snow and be able to drive to the ranch when she wanted to. She was angry at this gift and tried to pay Lonnie and Grace for the car, but they held fast, and she drove it proudly home.

When one of the horses that were a team broke his leg and had to be put down, Lonnie told Grace she could have whatever salvage money the horse brought. It was enough to build a counter bar in the newly-remodeled kitchen. This was the greatest! She could cook, turn around, and serve right onto the counter. Bar stools came soon after.

Grace's mom and dad, Nellie and Rich, helped furnish the house with used carpet and curtains, which had been retrieved from the local dump where he worked in Colorado, then cleaned. Rich proudly crowed that he had always told everyone that people were wasteful and threw things away that were perfectly good. The house turned out perfect.

The youngest daughter, Dar, started first grade at Wind River School on the Indian Reservation. Mickie was in third grade and Kail was in fifth grade. Kail complained that he got a cruel teacher who disciplined with "ear twisting"—that is what all the students called it, and it really brought great results and full cooperation. All the parents loved the teacher. Kail wondered if she had been a bronc buster in another life, since broncs always got "eared down" by the cowboys when they squealed and kicked.

Everybody in the Mantle family worked hard all week, but weekends were for fishing and rodeos. There was a youth rodeo organization in Wyoming. Lonnie built up a bucking horse string that he leased out for many of the youth shows. He also had a string for amateur rodeos. Lonnie taught the three kids how to compete in each of their events. Grace was willing to drive them and their horses to rodeos if Lonnie couldn't go, and they traveled every weekend to a kids' rodeo. They had fun, won lots of trophies, and made friends all over Wyoming. At those kids' rodeos everybody knew your name, your horse's name, your goat's name, your dog's name, and if your mom put out good food.

Lonnie claimed he quit riding rodeos when one year Kail got his calf rode in the kid calf-riding event. Lonnie had bucked off his bronc, and Kail swaggered up and asked him how come he rode his calf and Lonnie bucked off.

The Fourth of July for the Mantle family was always spent fishing. A pack trip was put together and they went far back into the wild Wind River Mountains to beautiful pristine lakes teeming with trout. The magical trip lasted for at least five days. This pack/fishing trip remains a yearly event, and even the grandkids get in on it now. One year they took Aunt Queeda on the ride. She had been very ill, and as the days went by, the fresh mountain air, the horse riding, and the fishing brought her back to life.

High school rodeos in Wyoming paid money, which helped with the expenses. All the kids helped pay their expenses with their winnings. Dar even went to college on a rodeo scholarship, in addition to her scholastic scholarship. All three of the Mantle kids rode on the college rodeo team.

Lonnie and Grace Mantle

❈ ❈ ❈

Lonnie and his family fondly remember John Moore, a neighbor from Hudson. He was a bricklayer by trade and built the fireplace in their home near Hudson, then came to Pavillion and built a huge fireplace in their new house. It opened on both sides and was the heat source for the remodeled house.

Later John and his family moved their trailer house out to the ranch in Pavillion, and he went to work for Lonnie full time. He was the only hired hand Lonnie needed to help move horses, irrigate, or whatever else needed to be done. John always used a bull whip when he worked livestock and was very proud of his expertise, but good as he was with that whip, he just could not get the cows to accept their newborn calves, no matter how many times he cracked his whip. The pastures where the livestock was kept always needed fences repaired. John told everybody he rebuilt all the fences in Wyoming and was starting on Nebraska.

John drove the semitrucks for Lonnie. One time in Cody, Wyoming, a highway patrolman pulled him over for going down hill too fast. He asked the patrolman where was he when he was going up the hill? It all averaged out didn't it? No ticket.

Lonnie's business grew until he had over 800 head of dude horses. He rented to resorts and stables and camps all over Wyoming.

※ ※ ※

Over the years, many youngsters found refuge from whatever their teen problems were at the Lonnie Mantle Ranch. Mark Nelson was having trouble in school in Jackson and was ready to drop out so he could rodeo. He came to stay with Lonnie and Grace so he could learn to ride saddle broncs. He got to stay on the condition that he stayed in school and graduated from high school with a C average or better grades. Mark was called Mick's and Dar's red-headed, left-handed stepbrother. He ended up finishing school, winning the state high school bronc riding award, and attending a two-year college. He now lives in Las Vegas, Nevada.

Kerrie Weeks showed up after a high school rodeo one weekend. She wanted to stay, and ended up living with the Mantles, finishing high school at Wind River. To make up for her missed work and get credit for a year of computer science, her teacher at Wind River helped her set up an elaborate bookkeeping program using the information from Lonnie and Grace's horse operation. It was a database program that they still use at the ranch. It is also used by Kail Mantle and his wife, Renee, for their horse-leasing business in Montana. Kerrie went on to college, held down jobs in bookkeeping, and now is the mother of two girls and lives in Denver, Colorado.

※ ※ ※

Kail lives near Three Forks, Montana, with his wife, Renee Daniels Mantle. They have a dude horse-leasing operation in Montana and Idaho, just like Lonnie has in Wyoming and Sombrero Ranches has in Colorado.

Mickie lives in Three Forks, Montana. She works as a bookkeeper for a few small businesses. She has her own sewing/embroidery business—"Bunkhouse Designs."

Dar married Bob Vogel, and they have always lived on the ranch near Lonnie and Grace. They have three children, Josee, Kit, and Kage. Bob and Dar recently purchased the horse-leasing business from Lonnie and Grace, along with its horses and equipment, and land. Although Lonnie still lives at the ranch, he now has the freedom to do some of the things he likes to do, like traveling, fishing, and just not working so hard.

The really sad news is that Grace Mantle passed away on August 25, 2007. Within only three weeks of revealing that she was sick, she was taken from her family. The whole extended family is mourning the loss of this person who gave so much life and love wherever she was.

27

Evelyn, 1970 to 1979

EVELYN WAS EXCEPTIONALLY SAD and introverted after the funeral of her son, Potch. It ended a lifetime of difficulty between them, beginning when he was a child. All the love she could give him hadn't changed him much, although he had mellowed toward her some over the years. She was glad he had married such a fine woman, and his daughters were her bright stars. She missed them terribly when they left. Her life suddenly felt pointless, so to give herself some enthusiasm, she took on paying off her house as her all-consuming passion.

The Christmas rush at work was upon her. She had too much to do, but Fashion Bar would not hire another lady to help. She brought work home to do at night, didn't take lunch breaks, and worked until the store closed late in the evening. The strain began to take its toll on her. She looked worn out all the time, and her eyes took on a sunken, desperate look. As often as she could come, Queeda invited her over to eat supper with them, as she knew Evelyn wasn't taking time to cook.

The Barnum and Bailey Circus came to town at the Denver Coliseum. One of the managers at work gave Evelyn tickets. She loved circuses and carnivals and wanted to treat her family to this fun she had so much enjoyed as a child. She took Cindy, Justin, Freda, Rex and Queeda to the circus. They screamed and giggled and clapped and played with the clowns through a glorious performance of the three-ring circus. Cindy asked, "Granger, can I come spend the night with you?" After she was in bed she said, "Granger, it's so nice of you to invite me over to sleep with you." Before Evelyn could even brush her teeth, Cindy was sound asleep.

That spring Eva and her friend, Genevive, visited from New York, but Evelyn was working so hard she didn't get much time with them. They worried about how tired she looked. They tried to help her out with a loan to pay off the house. She, of course, refused. She told them that she only had $900 yet to pay on her house, and by August would have it done. She was elated, because "Now she would have her own roof over her head."

In the fall of 1970, Evelyn felt like she could not bear it when Cindy left to go to school in Mexico. They were constant companions and enjoyed every second together. Evelyn wrote to Eva:

> *Cindy has left for her school year in Mexico. She is content so far. She is such an active girl and always doing something, while the Mexican women are rather inactive, and she is having trouble keeping herself busy. Before she left she would come over and we would sew like mad. I would cut out and she sew. Has lots of designing ability. I made Cammie's room into my sewing room, and am sewing new clothes and alterations—also my job.*

January 1971 began a very difficult year for Evelyn. Even with her new car, the icy roads and heavy traffic of Boulder to get to work were a daily drain on her. She had been overworked at the store since October. She missed Cindy, and the rest of the family seemed to be gone or in a mad scramble all the time.

At this time, in 1971, Evelyn wrote to her cousin:

> *I have numbness in arms and legs and passing out. Dr. put me on a new medicine that really did me in. I was a month getting back where I could do anything again. Dr. never did figure it out. They are too busy to really take care of anybody. But I really do have to slow down, and I hate it.*

She fondly remembered that Dr. Monahan in Craig had told her that if she could get word to him when she ever needed him, he would come to the ranch and care for her if he had to come by dogsled. Comparing his dedication and the obligation city doctors felt toward their patients frightened her.

Of course, she didn't tell her family about being sick. They should have noticed, but in their "busyness," they didn't. Queeda's family just kept too busy to be much company to Evelyn. Every Christmas vacation they spent in Tyler, Texas, with Rex's folks. Also, Rex had bought a horse-boarding barn and stable business in Scottsdale, Arizona. He was gone most of the time over several years tending this business, leaving his family to take care of everything at home. On January 29, 1971, Queeda, Rex, and the two little kids—Freda and Cody— left for Scottsdale in a pickup and camper, pulling a horse trailer. They would not be back until February 15. Cindy was in Mexico, and Justin stayed with his Granger. She loved having him, but it was more work in her already overstuffed days.

※ ※ ※

On February 25, 1971, Tim wrote to Evelyn from the Bower Place on Douglas Mountain. They had spent the winter there so they would not be trapped in the canyon when they needed out. Tim and LaRue were expecting a baby any time. Their baby girl was stillborn after a difficult birthing that almost took LaRue's life as well. The baby is buried beside Granddad Mantle and Charley Jr at Red Rock Ranch.

※ ※ ※

In the summer months, Queeda was working at the stables all week and hauling kids and horses to Little Britches rodeos every weekend. Evelyn felt like she was intruding when she took any of Queeda's time.

Lonnie and Grace were always very good to visit her during Christmas vacation, and they faithfully called her once a week to see how she was doing and tell her their news. They regularly sent her pictures of the kids. She never told any of the family that she was sick, though.

A few years later, an accident nearly took the life of Queeda and Rex's young son, Cody. On July 12, 1973, Evelyn wrote to Eva from St. Anthony Hospital in Denver:

> *Sunday, four-year-old Cody got kicked in the chest by a big packhorse and thrown down on the cement floor of the stable in Grand Lake. An ambulance came from Granby and took him to the airport in Granby where the Flight-For-Life helicopter from*

St. Anthony [hospital in Denver] picked him up. Too many people so Queeda couldn't go. His lungs were collapsed and they inflated them, and continually sucked out the blood clots in the air passages. The accident was at 3:00 p.m. and they had him at the hospital at 5:00. Miracle after miracle was performed by the doctors as they kept in constant conference with each other from the clinic in Granby, the helicopter, and the hospital. It was obvious to them they were losing him, so the doctor in Granby gave him a brand-new drug as a last resort, and it worked.

He has a skull fracture, but no blood clots or brain damage. He has a tube in his stomach pumping out blood. He is being fed intravenously. He was in and out of consciousness all night, but Monday he started walking a little and eating solid food. He still sleeps a lot. He is alert and talks and thinks well, so believe he will be OK.

Queeda drove back to Grand Lake from Granby, collected Freda and the baby-sitter. A cowboy named Dale Wilson who worked there drove them to Estes Park. From the stable there, Keith Hagler and Doyle Eaton drove them to my house, then took Queeda on to Denver.

I took the kids to Queeda's house to tend them where they would not be so upset. Rex was in the hay fields, and before he could be notified he had gone to Henderson to pick up Cindy and Justin from a Little Britches rodeo. Danny Souders found him, and traded vehicles so Rex could go right to the hospital and Danny brought the kids and horses home. Rex arrived at the hospital in time to meet Cody in the helicopter. Queeda thought Cody wasn't going to live, so she called their family preacher and he and the whole Church of Christ congregation were there to hold the family's hands. Rex called me every 15-20 minutes, so by the time Queeda arrived from Estes we knew Cody would live.

Cody was very anemic from losing so much blood. They opened his skull to allow for the swelling. His mother spent the night with him, as he was conscious enough to be scared all alone. He is black and blue. His skull will take two years to mend, but the doctors are confident he isn't going to suffer from any permanent damage.

Evelyn made time in her busy schedule to spend a lot of precious Grandma-love on Cody when he got out of the hospital. She sat quietly with him when nobody else could, took him for short walks, petted his dog with him from a blanket under a shade tree, and fixed him chicken soup and Oreo cookies.

Evelyn was elated when Cindy arrived home from Mexico. So what that she spoke perfect Spanish? It just hadn't been worth it to take that time from her life and the life of her family. She just wished this family would slow down and smell the roses. Cindy and her Granger spent three happy years together visiting each other often and doing projects together.

Cindy's last year at home was 1975. She graduated from Boulder High School in the spring of 1976. She was making a name for herself in high school rodeo by that time. In celebration of her success, her father had bought her a beautiful new horse trailer to pull behind her pickup, and she could drive to all the rodeos. She ended up in the top ten, which qualified her to go to the finals in New Orleans, Louisiana. She came down with the chicken pox, but wanted so bad to go to finals that her mother piled in with her and they drove the horses to Louisiana. Evelyn had never seen anything so crazy in her life, and she started worrying right away.

Her worries were well founded, but for a different reason. About 5:30 in the afternoon of July 30, 1976, atmospheric conditions combined to create massive death and destruction in Colorado. An ominous black cloud became stationary over the east side of Estes Park, right over the Sombrero Stable riding area. It dumped eight inches of rain in one hour. The Big Thompson River, usually a tiny flow, reaching from Estes Park into the flatlands at the mouth of the canyon and on into the little town of Loveland, became a raging devil, carrying a wall of water nineteen feet high. It threw around ten-foot high and bigger boulders like they were toys. There were estimates of 2,500 to 3,000 tourists in the narrow Big Thompson Canyon that night. The steep canyon walls left very little room for the road winding down through it. Road and campgrounds in the canyon were completely wiped away, along with buildings, campers, cars, and people. One hundred forty-five people died, including six who have never been found.

Evelyn was frantic. Justin, Freda, and Cody were all at their house in Glen Haven, the small town northeast of Estes Park. They were there with a babysitter. The radio and TV blared news that Glen Haven had many deaths and was totally destroyed by the flood. Of course, the phones were all down. Evelyn was pretty sure Justin had gone to Cheyenne Frontier Days with a date, and he would have been coming home about the time the flood hit. She tried to call Queeda in Louisiana, but the phone lines were jammed. Finally Queeda got through to her and told her that they were alright: Cindy was feeling better, and Rex had called, so he was alright. Soon Rex got through to her. He told her that Justin had been on the road, but he had not yet started up the canyon when the flood hit, so he was all right. The little kids were in their house, which sat high on a hillside in Glen Haven. Their babysitter was with them. They had not gone out to the Sombrero daily steak fry that night because the kids had come down with the chicken pox.

Although the water spout had let loose right on top of the steak fry group that was out that night, they had all gotten home safely. The water had been up to the horses' bellies in some places, but the people rode it out. So far Rex didn't know of any family or employees they had lost in the flood. He asked Evelyn if she would please man the phone at his house to tell people what had happened, that the family and Sombrero people were all alive, all the horses had been saved, but Highway 34 up the Big Thompson Canyon was completely destroyed. As could be expected, the rest of the year business at Sombrero Stable was almost nonexistent.

By September of 1976 Evelyn joked that her sewing was getting harder because styles were changing and everybody wanted her to make the mini-skirts longer. She missed Cindy, who was enrolled in Abilene Christian College in Abilene, Texas. She got reports from Grace and Lonnie that Aunt Nancy, Charley's sister, was with them—that she was not happy and was being bad to everyone. Evelyn thought she and Queeda might have to go to Wyoming and move her—maybe to California. Her sons lived in California, and Nancy was just not able to take care of herself anymore. In her frustration about her helplessness she had become impossible to live with.

The wild activity was still on at the Walker house, and Evelyn didn't get to see much of the family. Justin was playing football at Boulder High. He left at seven and didn't get home until around nine at night. Freda played soccer with a city league. Cody was only eight, but he had a full schedule playing soccer and was also in Cub Scouts.

Eva and Satie, Evelyn's cousins, drove in, picked her up under great protest, and went on a trip together. They traveled to the

West. Satie's daughters, Rena and Lucille, were their gracious hosts along the way. Evelyn was glad to get to meet this part of her family. It was a refreshing trip, but a tiring one. Evelyn was just not feeling well, and tired easily. She came back to so many eager customers that she had to turn many of them away. She was so glad that she had quit her job at Fashion Bar. At least she could sew just when she was able.

※ ※ ※

Along in October, hunting season opened on the Western Slope. Evelyn and Queeda had shopped and shipped groceries and supplies to all the hunting camps and were looking forward to a slower pace. Evelyn joked to her cousin:

> *Hunting has been hard to beat right here in the pasture out on the Rocky Flats south of town. Somebody shot one of Rex's buffalo. While dressing it out the whole herd attacked him and he had to leave it. The next week one night the buffalo herd decided to migrate East. They broke the fence down and while crossing the highway four were killed and four cars piled up as they came around the bend onto the accident scene. No one was hurt, but two cars were totaled. We couldn't contact Rex and Queed— they were at Justin's football game down in Denver—until after 10:30.*
>
> *The ranchers across from the pasture butchered the buffalo out, so saved the meat. Stan Johnson got in on it, and at his house they ground meat and stuffed sausage and partied until dawn broke.*

By 1978, Evelyn was very sick. She was determined not to show it, but she had a hard time doing the things she wanted to do. She took one last trip that summer with Queeda. They went to the Mantle Ranch, where she let the memories of her lifetime there play through her mind. She asked Queeda to take her to Hayden, where her high school class was having their reunion. She visited with old classmates, but she was sad that a boy she had been more than fond of, Jess, had not come. She traveled around with Queeda and the kids in the old semitruck to a couple of rodeos, then wanted to go home.

That fall, Evelyn had surgery to remove her spleen, and gradually her health deteriorated. On January 2, 1979, she passed away.

Her funeral was held in Boulder. There was standing room only for the huge crowd who gathered to say good-by to the woman who had touched so many lives in so many ways. Evelyn was buried in Boulder, Colorado.

28

Pat, 1967 to 1992

AFTER PAT MOVED TO THE WESTERN SLOPE of Colorado in 1967, when Sombrero bought the Big Gulch Ranch outside of Craig, he felt like he had come home. Here, a man who had been raised as the toughest kind of cowboy and knew all there was to know about ranching and livestock was respected. He had a lot of friends here: Everyone knew him and he was admired, and maybe even looked up to, as a "Real American Cowboy."

He had soon converted the upper floor of the old milk barn at the Big Gulch Ranch into living quarters. The long room on the west side was made into a dormitory with cots for horseshoers, wranglers and hunters. He had turned the stable in Grand Lake over to people he had trained in the business and who he trusted. Now he had to devote full time to the Western Slope part of his business. He and Rex started looking for an additional ranch on which to winter the fast-growing Sombrero horse herd.

In 1968, Sombrero bought a sprawling ranch in Browns Park, 50 miles west of Craig. It was strictly winter pasture country. It was going to be a tough place to manage, for many reasons. Water was scarce, so the horses would need a little snow to augment their water needs. The existing water sources were needing repairs to take full advantage of what water there was. It would take constant vigilance, moving the horses around to utilize all the feed. After some of the herd got used to the system, they would lead the new ones, and it would not take so much effort to keep them scattered.

One of the biggest problems was the great number of wild horses just north of the ranch in Sand Wash. The wild studs would come over and steal dude mares and claim them in their private herd. Not only did Sombrero not want foals, they didn't want to lose the summer

From left to right: Pat Mantle, Keith Hagler, and Rex Walker

work of the mares. Getting the mares back meant going out among the wild horses and rescuing the mares. Tough job, but somebody had to do it!

First, though, there was a lot of fence that was inadequate or completely missing around the borders of the ranch. Pat figured he had to get a crew in and build fences first thing. He manned a crew to cut cedar posts, bought miles of barbed wire, and got started on it right away. In the Mantle way, camps were set up at the farthest end of the job, and the fencing crew had to work back toward the ranch and finish the fence in order to earn a little time back in civilization.

Pat had set up his winter home in the house on the Curtis Place in Browns Park. It would make a good central location for taking care of the horse herd in the winter. The corrals needed work to make them good, but for now they were adequate. The fence crew would have a good comfortable place to live while they built fence. The turnoff from the highway to his ranch was marked by a white brassiere, size 40DD. It flew from high on a post.

Rex found a wino down on Larimer Street in Denver named Bob who claimed he was a pretty good cook and a damn good all-around ranch hand. Pat eagerly hired him and took him to Browns Park, where there was no way in the world to walk out to the liquor store.

Soon Bob sobered up and looked around and liked what he saw. Pat always gave a name to everybody, and Bob became "Mushy Bob." For many years he worked for Pat, cooking, feeding livestock, and doing odd jobs around whatever place they happened to be. Mushy Bob even set up a kitchen and fed the fencing crew out on the range, whether it be in a tent or under a cedar tree. His favorite place was hunting camp: hunters always brought lots of booze.

In March, they had to go round up the dude mares out of the wild herds. It was a tough job, but excitement surged through the men as they grained up their long-legged, toughest horses that could run top speed over rocks and brush and jump washes at a single bound. Cinches were replaced and latigos checked. New lariat ropes were stretched. All the conversations were about catching wild horses.

Elaborate plans were made and secret traps were built of dead trees and brush. Long widespread wings of brush were laid out, funneling gradually into the trap. The plan was to ride quietly toward a herd of horses, get them running all in the same direction, then riders on the sides would gradually crowd them into a controlled herd, all running toward the wings of the trap. The stallion was the leader, so it was most important that he didn't escape. Occasionally, a wise old wild mare would make a dash for the side and would have to be roped.

As the herd neared the wings, the trap was supposed to look totally innocent to the wild horses, because it was made of materials from the countryside. By this time, hopefully some old dude mare would take the lead and the wild bunch would all follow her into the trap, and the gate could be slammed behind them.

Usually that wasn't to be. The wild stud would smell a rat and bolt to the side or back through the bunch, causing complete panic, and the cowboys would lose the whole gather. At this point, each wrangler would usually take off through the sagebrush after a dude mare they had spotted, finally rope her, and contain her long enough for her to remember that other life she had as a tame horse, and she'd give in. Back they would go, leading their now-docile dude horses to camp. Stories were told for weeks about the chase and how the cowboys got whipped once again at their game by the wild studs. These stallions were admired and respected for their escapes, and it was no great source of embarrassment to the wild horse chasers that they got bested once again. There was always next time.

Pat was providing the livestock for many Western Slope rodeos with his 7-ll Rodeo string. He was always horseback in the arena as pickup man or helping with the livestock, so everybody knew who he was. Some men he knew from Steamboat approached him about putting on a small weekly rodeo there. They felt that not only would it be a fun thing, but it would set Steamboat apart as the one really "Western" ski area, as they wanted it to be known.

Pat had been thinking about opening a Sombrero Stable in Steamboat anyway, so it sounded like both together was a natural. He leased a location at a scenic old barn at the foot of Mount Werner for his stable. People could ride horses on the ski slopes, in summer, and trail rides could also wind up the deep timber-laden mountain valleys. He sought out a pretty little spot for his evening steak fry rides, and Pat's cookouts became famous in the area.

So now Pat was spending winters in Browns Park, summers in Steamboat, and spring and summer at Big Gulch. In spring, the horses were brought back from Browns Park to Big Gulch and prepared for their summer work. They got new shoes, cleaned up, sorted, and trucked out. In fall, this was the gathering place for all the horses. They were trucked in, shoes pulled off, wormed, and moved to Browns Park for the winter.

Pat had been raised on a cattle ranch located in rough canyon country with no road to it. Horses were used for everything. He knew what horses could do and he knew what men could do riding a good horse. It seemed perfectly reasonable to him to just set a date and drive the 600 head of horses for fifty miles to their winter pasture in Browns Park. It never occurred to him to haul the horses in trucks. This was the first of the soon-to-be famous yearly Sombrero Ranches Horse Drive.

In the *Steamboat Pilot @ Today's*, Rodeo Guide 2005, July 1, 2005, they ran a memorial issue on Pat Mantle:

> *Steamboat veterinarian Mike Goetchey worked at Sombrero Stables for Pat as a youngster. He remembers that Mantle had a nickname for everyone. That personal touch made people feel*

special, even the poor wrangler that Mantle dubbed "Baby Legs" after he turned up with raw, bloody knees following a tough ride. Mike also recalls his earliest memory of Mantle. At a jackpot rodeo in Boulder where Mantle was working as a pickup man, an ornery bull with a reputation for jumping the fence had thrown a cowboy in the arena. The bull, named "Long John," made straight for a spot on the fence where two little girls on ponies sat transfixed by the onrushing 2,000 pound animal. Just like in the movies, Pat threw his rope and caught Long John right at the top of his jump, and pulled him back into the arena.

The fall of 1969, the date was set for the first Sombrero horse drive. Former wranglers, local folks, old friends, and a few people he had never met showed up that morning for the "Biggest Event in Moffat County," as it was heralded. Some had their own saddles, others rode dude saddles, but all mounted up to be drovers on the biggest two-day horse drive anybody had ever been involved in. Pat looked over his crew and wondered how many of them would survive through the first rush of the horses out the long driveway and the sharp turn onto Highway 40.

The night before the drive was to begin, Pat had nominated a bunch of girls to be "string girls." Proudly they wore the name, and their job was to drive way ahead of the horses all along the route and tie flagged bailing twine across the cattle guards and shut any open gates so the horses wouldn't stampede off someplace they oughtn't to be. Pat himself and one passenger took the lead in Pat's pickup. Each had a red flag that they waved to warn motorists of the horses on the road. Another flagger drove behind the herd.

Mike Harding got the coveted job of opening the gate to begin the fifty-mile trip. Snorting and eager to run, the horses barged out of the corral and into the long driveway. The best riders were in front to slow down the charging horses so there wouldn't be massive slipping, sliding, and crashing when they hit the asphalt on Highway 40. Alongside the herd rode the next best riders: their job was to keep the herd from scattering. Behind the herd rode the drovers who were to use their best cowboy yells to move the herd forward and push any laggards while trying not to fall off their horses. The roar of clattering hooves was beautiful, and maybe a little frightening.

All day the herd pounded down the highway past gray sagebrush, rolling hills, and cedar-tree crowned mesas. A noon stop and short rest while Pat roped replacement mounts for everyone was welcomed.

The only food and drink came from whoever had brought something in their pocket. Finally, late in the afternoon, they passed down the main street, which is Highway 40, of the little town of Maybell. The whole town turned out to watch and wave and shout greetings. Soon after, the herd turned right onto Highway 318. That first night, the horses were put in a pasture, fed hay, and rested up. The riders went back to the ranch for the night and Mushy Bob fed them a big supper. They went upstairs and threw their bed rolls on cots or on the floor or slept outside under the stars. The second day, the drive started again at daylight after Pat roped a fresh mount for each drover. Sore butts and high spirits were the order of the day. Everyone was glad when late in the day, the last horse finally trotted through the gate into the winter pasture. All the riders felt a sense of accomplishment that warmed their souls. The ranch hands stayed to scatter the horses over the ranch the next day and the weekend cowboys went home. The first of what was to become a great Sombrero tradition had come and gone.

Pat didn't set many rules for people to follow when they came to his horse drive. However, the one hard and fast rule was—no women and dogs. It was 1976 before the first woman was allowed to ride a horse and participate. That was Maggie Bentz from Steamboat. She could outride most of the men. Shortly thereafter Peggy Miles, wife of John Miles, was also allowed to ride with the men. Over the years Pat allowed several women to join in who could "make a hand and not be in the way." They became his loyal fans, and he bragged on their skills. The rule on no dogs remained. If you brought your dog and he caused any trouble he was likely to be shot.

Pat became well known for his terse answers and Western wit. People who got too nosey were often the target. He once told a newspaper reporter, "When we were kids, if we told our daddy we were hungry, he'd hand us a stick and point at a jackrabbit." This kind of story and hundreds more followed Pat all through his life.

Over the years some winters in Brown's Park were terrible. Snow was so deep and constantly fell that the horses had to be gathered and fed with snowmobiles. Pat's most faithful winter cowboys were Mike Harding, John Miles, and—later on—his own son Steve. Mike and John were from Peoria, Illinois, and had come West to be cowboys. They soaked up Pat's ways and knowledge like sponges and put their very lives in danger doing all the things he taught them to do. They got so they could handle the difficult horses, survive any kind of weather, cook, drink, and cuss.

All that hard work was sweetened by the wild horse chases in the spring. How they loved the intense planning and execution of capturing a herd of wild horses and stealing back the dude mares! It was a wonder they got through it alive.

※ ※ ※

Pat was busy with his stable and his rodeo string all summer. The weekly jackpot rodeo in Steamboat was a big success and an asset to the tourist business there. It gave anybody that wanted to the chance to ride a bronc or a bull, or to rope, or just sit and watch the action.

One winter, Pat bought a bunch of snowmobiles and went into the snowmobile rental business. Tim joined him in the business a short time later. It was going good until one winter, there was just too much new snow every day for the snowmobiles, and they sat idle most of the time. It was a good time to make needed repairs. One day, they were doing snowmobile repairs inside and the whole building burst into flame. That was the end of the snowmobile business.

※ ※ ※

The following Thanksgiving story is paraphrased from "A Thanksgiving horse drive to Colorado's winter range. Hoofbeats into Yesterday," by David Thompson. It appeared in *The Colorado Magazine—Colorful Rocky Mountain West* in the Nov/Dec, 1973, issue.

The day before Thanksgiving, 1973, at the Sombrero Ranches headquarters at Big Gulch, excitement was running high. The 7,000-acre horse ranch was being gathered. Stable horses, camp horses, hunting horses, and rodeo horses had been turned out here at the end of the tourist and hunting season. Pat Mantle owned this ranch and these horses with his partner, Rex Walker. The gather would be made of the pasture in preparation for the drive to begin on Saturday.

Pat was standing on a hill watching his crew gather the horses. Mantle said, "This is the biggest horse drive in five states—the last of the big ones. Nowadays most horses are trucked when they're moved." He scanned the expanse of his ranch. "The wild Mexican is down there somewhere. We call him that. His name's Mike, up from Durango, Mexico. Can't hardly speak a word of English, but one hell of a cowboy. Today he's ridin' a rodeo bronc, and I imagine that old horse is learnin' somethin'." Pat pointed out where his foreman,

Barney, was driving horses off a hill three miles away. They got all but two gathered and into the corrals by Thanksgiving eve.

On Thanksgiving day the last two horses were gathered. A huge Thanksgiving dinner was laid out by the ranch cooks, and ranch hands and visitors ate their fill and ate again. By this time, many of the riders for the drive were arriving, and on Friday the rest got in: Earl Vaughn, a 49-year-old ranch hand, uranium miner, and oil rigger; Tim Mantle; Mike Harding, in his twenties, a friend and part-time ranch hand of Pat's; a carload of Pat's friends, including the Gruber brothers, Jerry and Don, construction contractors in Denver; Annie, Don's twenty-five-year-old daughter; and a pack of college-age kids with new hats and boots.

A snowstorm set in in earnest, delaying the 8:00 a.m. start. Pat would not put his horses in danger in the fast-moving traffic, slick asphalt on Highway 40, so they waited for a change in the weather. Finally, the snow let up and they could go. Pat took the front in his pickup to flag oncoming traffic. Methodically, they turned the horses out of the corrals and forced them into a long line, reaching over a mile from front to back. When they got to the highway, they were well under control, not allowed to run, or to bunch up.

A new Buick station wagon, with chains popping on the pavement, tried to work its way through the herd. They would probably have been there the entire day if Tim had not ridden up beside the driver and told them to force their way on through.

The herd rounded a bend and quickly overtook a herd of sheep on the road, also going to winter pasture. Before anyone really knew what happened, the lead horses were in among them. The sheep panicked and began stampeding in all directions. Then, to compound the problem, a huge horn-blasting semi-trailer truck began pushing through the whole mix-up. Pat came riding back to supervise. They crowded the horses to the middle of the road, while the sheep were separated into a tight circle to the right of the pavement by the herders and their amazingly quick dogs. When the horses were reorganized along the road, they were urged slowly forward, past the circle of bleating wool. Meanwhile the semi and several other vehicles edged by and the chaos was over.

The Lay-Route post office appeared on the left. It consisted of a post office with two gas pumps in front of it. Tim explained that before the turn of the century, ox and cattle trails crossed here. Where the post office now stood was a saloon, manned by a regular squad of girls. One night a bunch of dusty bullwhackers and the hot-eyed cowboys rode in at the same time, and a bloody fight ensued over first rights to the ladies. The next morning four bodies were carried up the hill across the road: three bullwhackers and a cowboy. The weathered grave markers are still visible.

Lunchtime came at a steel suspension bridge across the Yampa River, just east of Maybell. Sandwich fixings and beer were available. After they ate and rested a short time, Pat roped fresh mounts from the herd and everybody saddled up. Crossing the bridge, the horses were packed tight from side to side, heads high and ears twitching as the hooves of 600 horses on the bridge shook the girders and made a horrendous sound, like deep thunder.

Passing through Maybell, the usual boisterous crowd came out to wave and shout their welcome. Soon afterward, the herd turned right onto Highway 318. On down the road, they put the herd in a pasture at Harley Guess's ranch for the night. Next morning the tired riders were on their way.

The drive wended its way along the foot of Douglas Mountain with not much excitement. Suddenly there was a subtle change in the horses: in the distance was a steel suspension bridge. The horses speeded up, and nickering and whinnying were heard throughout the string of horses. The men in front were hard-pressed to keep the horses from bolting past them. There were some tired horses plodding along at the best rate they could, and the herd had to move at their rate.

Not far from there was the gate into the winter pastures. These tired dude horses looked forward to grazing on native grasses with the freedom to roam and be "wild" that these pastures gave them. Pat watched his horses fondly as they wandered off to enjoy their "vacation."

Pat died in March 1992. He was visiting dear friends in Oregon when a massive heart attack took his life. His son, Steve, picked up Pat's casket in Denver and drove him to Craig. He was buried on his Browns Park Ranch. A long mournful crowd trailed up the hill behind a team pulling a wagon with the casket on it. Pat's favorite horse followed slowly to the burial sight. The horse was saddled, and Pat's boots were set into the stirrups backward.

29

Mantle Ranch vs. United States Park Service, et al

THE AUTHOR EXPRESSES HER SINCERE GRATITUDE to Lonnie Mantle, who contributed this chapter. Lonnie recorded many of his observations leading up to and during the trial, and they are paraphrased below.

When Dinosaur was scheduled to again expand and also administer grazing rights in 1960, landowners, ranchers, and miners demanded some assurance that their livelihoods would be protected. After all, the National Park Service had just proclaimed that grazing was an incompatible use. In response, Colorado congressman Wayne Aspinall sponsored public law 86-729 that tied grazing rights to private property inside the Monument indefinitely. Dinosaur National Monument grazing rights tied to private property outside the Monument would terminate in twenty-five years [1985], with the exception that those rights would continue during the lifetime of the original permitee and his heirs. Those grazing rights could now only be terminated if the federal government bought the ranch, or if a violation of the terms of the grazing permit occurred.

Thus began forty years of concentrated federal government action to put the Mantle family out of business by whatever legal or illegal methods they could devise. The Hells Canyon (Castle Park) 160 acres was the only vehicular access to the Yampa River Canyon area for thirty miles. This would be an ideal place for a ranger station if the National Park Service could gain control or ownership of it.

There were already hundreds of boaters a year floating the Yampa River past the Mantle Ranch. In the minds of the newly-appointed DNM administrators, the Mantle Ranch presented a threat to the management of the Monument. The Mantles lived in the heart of the Monument, had "trespass" rights on 30,000 acres of park lands, and had special privileges that administrators had no control over. There were increasing conflicts with boaters camping in the Mantles' hay fields and boat concessionaires trying to launch or retrieve their boats and guests on the Mantles' private land.

One of the largest concentrations of prehistoric Indian petroglyphs in the West, known as Hells Panel, on Castle Rock, was a favorite stop for boaters and tourists. To protect their personal property, as well as discourage the increasing damage to the archeological artifacts, the Mantles were forced to post their property and accept only invited visitors and friends. On the Yampa Bench grazing allotment, boaters and road tourists complained to park officials of cattle in the river corridor near camping spots, and roaming freely along the Yampa Bench road. Many of the park personnel were from the East and were also astounded to see ranching paraphernalia on NPS lands.

Tim and his wife, LaRue, along with Evelyn, remained on the home ranch at the time DNM began administering the grazing in Dinosaur. It was becoming very obvious to the Mantles that "Big Brother" was taking over, and that they needed to prepare a survival strategy. Special Use Permits (SUPS) were permits issued by the Park Service to designate what facilities and privileges the Mantles were entitled to on their grazing permit. This included reservoir repair, cleaning of springs, etc. The SUPS were valid for ten years, at which time they were to be renewed or be replaced by an Allotment Management Plan (AMP). After 1975, these SUPS were never renewed by the National Park Service until a Federal judge ordered them valid in 1997.

Dinosaur National Monument Superintendent Semmingson had prepared the SUPS and changes in the grazing livestock numbers for Evelyn and Tim to review at 9:00 a.m., August 6, 1964. Both parties agreed to the wording and agreed to return in the afternoon to sign the final document. At the time of signing that afternoon, both Evelyn and Tim noted new wording added that would further reduce their legal grazing capacity. The Mantles refused to sign and were given a take-it-or-lose-it option by the superintendent. Mantles were forced to hire an attorney to restore the lost grazing privileges.

At the same time, a new survey determined that the Mantle residence *was not on the original homestead*. This incident is related in Chapter 19.

This—1964—became the year for DNM to show even greater determination to reduce the Mantle Ranch's profitability or break their will to continue. Unbeknown to the Mantles, DNM had slated the building of new roads into the canyon country, and an appraisal of the inholdings of Mantles had been done by the NPS. Interoffice memos made note that the Mantle properties needed to be acquired before the new roads were built, because the roads would enhance the value of the land. In a memo from Superintendent Semmingson to the regional director obtained by the Mantles many years later through Fredom of Information Act (FOIA) requests, Semmingson wrote, "When land acquisition funds become available have a second appraisal made. In the event that the owner refuses to sell the lands to the government at the offering price, it is recommended that the property be acquired by condemnation with a declaration of taking. Also that the Castle Park (Hells Canyon) property be acquired for development." This undoubtedly meant the building of the NPS-coveted ranger station. The other acquisition was the Red Rock property, which was slated for new roads.

Condemnation loomed in front of the Mantles for the next forty years. To successfully condemn private property, the government must have a current appraisal and funds available. However, a condemnation could not occur until the willing buyer-willing seller clause had been satisfied. Using this loophole, the Mantles were able to keep the NPS at bay. The family would encounter at least six more appraisals in the next forty years, several of which were done by the NPS without the Mantles' knowledge or permission.

Superintendent Semmingson and his administrators apparently felt they needed more control over the grazing and actions of the Mantles and their visitors. A fire lookout sentry was located in a tower with a view of all the Mantle's private and range lands. The sentry had instructions to log all Mantle ranch activities. A powerful telescope would document license plate numbers of vehicles and the movement of livestock and riders.

Tim and LaRue stayed on at the ranch, which by now had been reduced to a ranch that would barely support one family.

In 1973, District Ranger Hal Greenlee, in a letter to the superintendent, recommended that a registered letter be sent to Tim to demand that he answer charges, such as extreme overuse of his grazing allotment, unauthorized use of a bulldozer, the use of a proposed wilderness area for grazing, unauthorized operation of vehicles off established roads, and unauthorized snow machine use for feeding livestock. This was a prime example of new or replacement park personnel who either had not read or had disregarded the 1964 SUPS, as well as the Aspinall Congressional Act authorizing all of the allegations.

About this time, DNM appointed Resource Management Specialist Steven Petersburg, who—for the next twenty-five years—implemented a strategy that would force the Mantle Ranch out of business: it would be necessary to create a trespass situation, obtain a conviction and terminate congressionally-mandated grazing. This would leave two almost valueless pieces of isolated property in the Monument with a questionable access problem. By 1976, Petersburg had divided the ranch into five separate pastures *with no fences to separate them*. When questioned about these new pastures by Tim, Petersburg's response was that they were defined by "natural boundaries." Grazing permit numbers were slashed with pencil reductions, and Tim's option was to sign the permit as written or lose it.

A new superintendent, Cecil Lewis, was sympathetic to the Mantles' plight. He cooperated with Tim in his livestock operation to the extent that he renewed the 1964 SUPS in 1975, and even added the right for Tim to feed his livestock on Monument lands with motorized vehicles and snowmobiles. Petersburg was forced to retreat till a new superintendent more sympathetic to his trespass plan took over DNM.

In 1981, new Superintendent Joe Kennedy took over the reins at Dinosaur. Kennedy sent Tim a letter in October notifying him that the NPS, in cooperation with the Colorado Division of Wildlife, was transplanting 50 to 100 antelope on Red Rock. Ten days before this letter was sent, Petersburg, in a handwritten memo to Kennedy, recommended "introducing antelope on the range which should show an overgrazing problem." Later, when the Mantles filed suit against the National Park Service, Petersburg was forced to read his own memo while on the witness stand.

In 1982, Kennedy notified Tim that *a new survey* of the Mantles' inholdings would take place. Condemnation was again on the minds of the Mantle family. Memos and letters during the next few years gained through FOIA requests by the family showed attempts to restrict grazing on the Yampa River corridor, along with refusal to allow mechanized repairs to maintain improvements listed on the now-expired 1975 SUPS. To legally comply with their own grazing handbooks, these SUPS were supposed to be replaced with an allotment management plan. DNM refused three AMPS submitted by the Mantles, and even though a Federal judge later ordered DNM to complete one, they disregarded his order.

A drought period began in 1983. It lasted several years and was compounded by the influx of Mormon crickets, which devoured everything. Tim was forced to retreat to his ranch at Meeker with most of the livestock, taking nonuse on much of the Mantle Ranch permit. Petersburg and his superiors seized this opportunity to restrict seasonal road use and even access to the ranch. Although Petersburg tried to prohibit feeding livestock on the private land, Tim was forced to ignore many of these attempts to restrict or eliminate everyday ranching activities. Documentation of livestock sightings, with times and dates in specific "pastures," became a priority for all DNM personnel. Heated confrontations with trespassing park personnel became common.

In 1988, Superintendent Dennis Huffman took over leadership of DNM. He seemed to be the man Resource Management Specialist Steven Petersburg had been waiting for. Huffman apparently vigorously agreed with Petersburg that the Mantles needed to be removed from DNM by whatever and all avenues available.

In 1981, DNM had entered into an agreement with Moffat County to help maintain the Yampa Bench road. Moffat County would later number and claim the Yampa Bench road, thus ensuring the Mantles access to their properties. Mantles owned the only private property on the Yampa River and Yampa Bench for forty miles. After the twenty-five year Aspinall bill time frame of 1985 phased out all other grazing permit holders, the Mantles held the only grazing rights inside the Monument south of the Yampa River. They controlled the only springs in the Red Rock area and the only vehicle access to the Yampa River in the whole of the Dinosaur National Monument Canyon area. The Mantles knew now that this remote property was getting more valuable every day, with or without the grazing rights. They also knew that 30,000 acres of grazing/trespass rights was a huge bargaining tool with the National Park Service.

Huffman began a barrage of charges against the family, such as livestock trespass allegations, illegal use of snowmobiles for access as well as feeding, off-road use of vehicles for maintenance of improvements listed in previous SUPS, and numerous other everyday ranching activities. He used the drought as a tool for a reduction of livestock numbers. Permits were now going to be required for all activities, and DNM virtually demanded access to the Yampa River over the Mantles' private property. *Huffman then closed the west road on December 5, 1988, thus locking the Mantles in or out of their ranch.*

Longtime family attorney and ex-FBI agent Stan Johnson advised the family that, given the apparent animosity between the NPS and the Mantles, he would recommend several precautions: don't travel into the Monument alone; carry a camera; tape record conversations with DNM employees; don't carry any guns in your vehicles; and refer all DNM-generated documents and encounters to him. Stan notified park officials of his spokesperson authority of the Mantles, and that they were to send all further mailed documents to him.

On June 3, 1991, Petersburg issued a memorandum to all Monument employees: "Monitor use on Mantle Allotment. Count cattle in each (pasture), record times, dates, brands, colors, breeds, etc." A handwritten note stated: "This is a high priority need. Every effort should be made to support it."

Red Rock Canyon was the historic trail between the Yampa River and the Red Rock Ranch, as well as to Chews' old home and west Blue Mountain. It was also the main artery of livestock travel between Castle Park and Red Rock Ranch. Huffman hired an archeologist named Jim Truesdale and a botanist from Colorado University to document "archeological sensitive areas" and endangered plants that might live in this area. Using old archeological records from the thirties and forties, Truesdale found plenty of archeological history, and the botanist, Tamara Nauman, documented a threatened bog orchid plant living in Red Rock Canyon. Both Truesdale and Nauman held credentials that would certify them as experts in a court. On October 10, 1992, Huffman declared Red Rock Canyon closed to trail and livestock use. *He had just cut the ranch in two.* Although there were already some existing fences, DNM constructed two new ones at each end of the canyon. Game and livestock continued to use existing side trails and riverside entrances to enter the canyon. However, on paper, the closure looked good to government officials and gave them more ammunition for more trespass action against livestock.

In February 1993, Tim and his son, Dean, stopped at Dinosaur headquarters to advise them that they were going to the ranch to check livestock and the general ranch status. Petersburg and chief ranger Nick Eason refused them access; the roads were closed for the season!! Tim and Dean went anyway, not knowing if they would be arrested. They found several cattle at the mouth of Red Rock

Canyon, trapped behind a gateless fence the Dinosaur personnel had built when it closed the canyon in October. Two head of cattle had already starved to death, and the others were in poor condition.

By the summer of 1993, Huffman and his administrators were now ready to deliver the blow that would destroy the Mantle Ranch. On August 2, 1993, Huffman sent a registered letter to Tim, notifying him that trespass action was being taken against him and that such action could result in revocation of his grazing permit. On August 25, a follow-up letter was sent to each of the five heirs of the Mantle Ranch. He stated in each letter: "We urge you to take immediate action to rectify this trespass fee situation. If you have no interest in this situation please notify us of the same by letter at the above address." In a letter to U.S. attorney Jerry Cooper, Huffman stated, "We expect other family members to pressure Tim into paying the trespass charges." This would pave the way for a criminal trespass and subsequent cancellation of all grazing.

Mr. Huffman received notification from each and every heir that they certainly were interested in this situation, plus several more pages of documented past harassment, trespass allegations, road access refusal, Red Rock Canyon closure by using the endangered species law, refusal to rewrite the 1975 SUPS, refusal to implement an allotment management plan, trespass on the Mantles' private property, a bill of $3,000 dollars for the dead cattle trapped behind his gateless fence, the demand to not feed on federal, as well as private property, refusal of access to ranch, home, and improvements on the allotment and the ability to maintain them, his implementation of livestock removal because of drought conditions without scientific data, the creation of a series of unfenced "pastures" to create this trespass situation, the fourteen years it took for the government to correct the patented land status in Castle Park, refusal of access to friends, relatives, or business associates, gestapo-like surveillance, and the general treatment of the Mantle family as if they were criminals!!

The family followed these letters to Huffman with individual letters to their political representatives, asking for help and/or censure of the National Park Service. Follow-up letters from congressmen and senators from Wyoming, Colorado, Utah, and Wyoming poured into DNM headquarters, demanding answers to all the allegations the Mantle family had claimed. Local newspapers picked up the story, which went on the national wire. Many Western livestock papers as well picked up the story and followed it for the next ten years.

Next, the Mantles started sending Freedom of Information Act Letters to DNM headquarters, asking for all documents pertaining to every aspect of the past thirty-three years of harassment. The most important FOIA was obtained by Tim. Armed with a legal document authorizing him to scan the DNM files, he copied interoffice memos, letters to NPS regional officials, and many other documents portraying the NPS's plan to destroy the Mantle Ranch and subsequently gain ownership of it.

Attorney Stan Johnson agreed to oversee Tim's pending case, but he also advised the family that to go on the offensive, they would need to eventually file suit against the NPS. The Mantles were agreeable to this and hired the Budd-Falen Law Offices, L.L.C., located in Cheyenne, Wyoming, as their defenders for the trespass action. At the first meeting between the Mantles, their attorneys, and the Federal attorneys, the latter portrayed the Mantle family as anti-government activists.

Tim's trespass fine was assessed at $462 and, of course, he had no intention of pleading guilty to the charges. Administrators at DNM were in constant contact with their regional office solicitors—the BLM, Washington, D.C., and FOIA refusal expert, Jack O'Brien, at regional. Tim's charges were in the criminal category. In a letter to Thomas O'Rourke, criminal division of the U.S. Department of Justice, dated December 29, 1993, U.S. attorney Jerry Cooper stated that DNM officials wanted the following actions: to levy trespass fees; criminal action; permit modification; impound cattle; no new grazing permit be issued till trespass fees are paid (this would put all of the Mantles' cattle on the allotment into trespass); and finally, termination/suspension/revocation of permits for continuing non-compliance. Huffman/Petersburg and associates were at last asking for all they had been planning for so many years. With the mounting political pressure and broad news coverage of the case, along with all the copied letters and memos, the regional office advised DNM to proceed with caution. They reasoned that if Tim paid the trespass fee, no more action was necessary to immediately take the grazing permit.

After much legal maneuvering, Stan advised Tim to pay the $462 "Under Protest" and to apply for the 1994 grazing permit and pay for it with a separate check. He also advised the Mantles to proceed with legal action against the NPS and to file civil charges against Superintendent Huffman.

The Yampa River corridor suddenly appeared in DNM letters as *a new "pasture"* with no use authorization! The permit application submitted to DNM in 1994 reflected use for forty-three head of cattle, compared to the approximate 500 head it had historically run for the first fifty years of the Mantle Ranch's existence.

On June 28, 1994, Stan Johnson, Tim, Queeda, and Lonnie met with Regional Director Ron Everhart, Bob Moon, and Bob Reynolds to discuss the expectations of the NPS and vent the Mantles' frustrations with the existing NPS' attitude and policy. This mediation meeting was a legal requirement [to try to solve the dispute] before further court proceedings could occur. The NPS demands and ultimatums were as follows: NPS will not open Red Rock Canyon, owns all fences and improvements, will not allow feeding by snowmobile, requires a Mantle representative on the ranch every day for daily communication with NPS personnel (there are no phones or electricity at the ranch), access would be allowed by Tim being furnished his own lock and key, with complete uncertainty by the NPS as to who was responsible for the fences, their condition, or purpose. Stan gave the NPS till July 20 to give a written summary on their position on all questions posed at the meeting. He also told them that should their answers not resolve the existing conditions in operating the ranch, he had the authority to proceed to the court system. In their report to Stan on July 19, the NPS held firm on all of their restrictions, and even ordered the Mantles to furnish them with access across private property both in the Monument and in adjacent land outside the Monument before they would issue a lock and key to the Mantles to gain access to their ranch!

Karen Budd, one of the family's attorneys, filed a twelve page lawsuit against the NPS on the Mantles' behalf in the U.S. District Court of Wyoming on November 21, 1994. The case was later moved to District Court in Denver, Colorado. Named as defendants were Bruce Babbitt, Secretary of the Interior; Roger Kennedy, Director of the NPS; Robert Baker, Regional Director; and Dennis Huffman, Superintendent of DNM.

By 1995, with mounting legal expenses, it became obvious to the family that they were going to need money to fund the lawsuit. Queeda contacted the Moffat County planning board and started the process of establishing various commercial businesses on both parcels of land inside the Monument. Castle Park and Red Rock had the potential for campgrounds, a store with sale items that would accommodate travelers by vehicle or boat, a hang gliding business off the 1000-foot cliffs, an ATV race track, and a helio/airport.

When the planning board presented their final request to the county commissioners for their vote, Superintendent Huffman spoke on behalf of the NPS. As the only neighbor to this property, he should have been notified, and, therefore, the county commissioners couldn't consider a vote on the request. One of the commissioners produced the signed certified receipt for the notification letter Huffman had received. The commissioners passed the commercial applications for both parcels by a unanimous vote.

In October of 1995, the NPS ordered *yet another survey of the Mantle property*, this time by the U.S. Geological Survey. The survey done in 1982 by the NPS was not valid in court. Again, condemnation loomed as a probable cause for this survey to be done. [The 1982 survey had showed two of the Red Rock Springs to be several feet off the private property.]

Red Rock had four very good springs that fed the irrigation system Tom Blevins used to farm his homestead during the early part of the 1900s. The springs, adjudicated on maps and noted as Tom Springs 1 and 2, were at the foot of Blue Mountain and at an elevation that would gravity flow water to the total 360 acres. With a storage facility, it would support not only the campground, but an entire subdivision, should future plans warrant one. DNM personnel trespassed at will to gain access to the land, and the Mantles filed trespass charges against the survey team, as well as the DNM employees. When the U.S. geological survey team and the DNM guides saw that Tom Spring 1 and 2 were surveyed off the private property, a plan evidently began in their minds as to how they could discourage the commercial use Moffat County had approved.

In June of 1996, by helicopter and hiking through the rough terrain around the private property, a team of NPS employees descended on the conveyance structures used to collect and store the water from the Red Rock Springs. Using axes, hammers, and other hand tools, they destroyed pipes, cement abutments, and rock work that controlled the flow of the water. As their final act, they stuffed the underground pipes with rags and bentonite so water could never flow through them again. In Tom Spring 1, Tom Blevins had driven a pipe in the rock wall so that the smaller spring flowed out of the pipe. This ninety-year-old landmark was also beaten out of the rock wall and deposited in the pile of rubble from Tom Spring 2. This was the most deliberate, obvious, and heinous act the NPS had committed in their quest to destroy the Mantle Ranch.

Queeda's daughter, Cindy Walker Lisco began a detailed accounting of the Blevins homestead, delving into Colorado Water Law. The Mantles hired engineer Bill Van Horn to do the percolation tests on the ranch for the commercial development. Bill was also very helpful with his expertise on Colorado, as well as Federal water laws. The family hired attorney John Henderson as a water law expert. Much research by Cindy Lisco and John Henderson's law firm determined that even though the water did not originate on the original homestead, it belonged to the landowner and he could divert it onto his land.

Previously, in March of 1995, Lonnie had hired a range use expert with superior credentials, Dennis Phillippi of Bozeman, Montana, and his business, called "Natural Resource Options Company." In July, he and his two sons spent several days documenting random plots on the Mantle allotment, as well as on the private property. He kept up this meticulous documentation for thirteen years.

During these years, grazing permits from DNM were due in February, but always late by at least sixty days. In May 1996, Lonnie received a trespass violation in the mail from DNM for trailing cattle to the allotment without a permit. DNM had failed to process the requested permit before the movement time. Finally, ten minutes before trial time, the U.S. Attorneys dropped the charges to a case they could not possibly have won.

Stan Johnson, attorney, rodeo announcer, photographer, and friend

On August 8, 1996, the NPS requested a Temporary Restraining Order (TRO) against the Mantles, claiming "irreparable harm and threatened irreparable harm to public and natural resources on NPS lands."

The old fox, Stan Johnson, warned the other Mantle attorneys not to consider this just an insignificant hearing on a TRO. He said, "This IS the trial. Contact all witnesses, prepare all legalities, and be prepared as if this was the main trial." The date of the hearing was set for October 4 at the Denver Federal Courthouse.

The diversion of the now-destroyed Tom Springs 1 and 2 was addressed by lawyer John Henderson, with Colorado Water Law pertaining to its use on homestead lands. Homestead and Colorado law gave ownership of that water used on the homestead to the landowner, with the right to divert that water even if it did not originate on the homestead, if the owner could prove it was used by the landowner to "prove up" on his land. Enola Chew Burdick was a ten-year-old girl when Tom Blevins broke his leg in 1912 and rode his horse four miles to the Chew ranch for help. Enola and her sisters walked each day to Tom's cabin and took care of his animals, as well as watering his garden, orchard, and crops while he recovered. When lawyer John called ninety-year-old Enola to the stand, everyone in the courtroom was fascinated with her detailed accounting of not only the diversion and use of the water, but her accounting and recollections of living during that era.

From left to right: Queeda Walker, Enola Chew Burdick, Tim Mantle, and Lonnie Mantle

Dennis Huffman, Superintendent Dinosaur National Monument standing beside the Red Rock Springs destruction on Judge's tour

After three days of testimony by numerous witnesses, Judge Kane made his own surprise announcement. He wanted to see all the areas of concern personally with his courtroom staff present. After his viewing, he would render a decision. The U.S. attorneys objected, because many of these areas were several miles from the nearest road. The Mantles immediately offered to furnish horses for the entire court, lawyers, the NPS, news media, and any witnesses needed. The offer was accepted by the court.

Friday, the eleventh of October, was selected as the date. Judge Kane arrived at the prearranged meeting place at DNM headquarters. Horses and riders were transported to Red Rock Ranch, where everybody mounted up.

The first stop was the Red Rock Springs diversion and the NPS demolition issue. Judge Kane inspected the site, both springs, and the rubble, with Tim explaining the off private property and diversion procedures. Old ditches, the storage pond, and farmed land were obvious. Judge Kane walked to the rubble pile, picked up the pipe Blevins had driven in the rock wall for Tom Spring 1, and stated: "This is lead pipe!"

Steven Petersburg, Resource Management Specialist at Dinosaur National Monument, holding down the wires in the gateless fence he had authorized at the mouth of Red Rock Canyon so the court could ride across.

Next was a horseback ride through Red Rock Canyon with discussions at intervals about endangered plants and archeological-sensitive areas. No DNM personnel knew where the plants were. Tim knew where the archeological sites were and pointed them out. They were in caves, most of which were inaccessible to livestock.

At the mouth of the canyon Tim called Steven Petersburg to let the party through the gateless fence DNM had constructed and where the two starved cattle were found in 1992. While Petersburg was struggling to take a portion of the gateless fence down, Judge Kane was taken to the adjacent Buck Cave to view the ancient deer petroglyphs on its walls, as well as the debris left by NPS personnel containing wire, posts, boxes, and plastic wrappers.

Crossing a portion of the fence wire being held down by Petersburg, the party proceeded downriver and forded the Yampa to Laddie Park. Here, they viewed the fifteen-acre park and the campsite where approximately 3,000 visitors had camped during boating season. This was the area where Huffman/Petersburg had claimed overgrazing. About fifteen head of livestock had used this park during the past year, and the nonuse by livestock was obvious. The extreme use by boaters was also very obvious.

The party then proceeded on to the Castle Park property, where many of the party picked and ate apples as they rode through the orchard and hay meadows. They then crossed the private land to the dump site where the snowmobiles, plastic water containers, and various other items were stored by the Mantles on DNM land. After a short look, everyone proceeded to the ranch house for refreshments and a drink of the spring water flowing via plastic pipe by the house. Here, horses were left and vehicles were waiting.

At the head of Hells Canyon, all court personnel got out and walked 300 yards on a well-used trail out of the canyon (Yampa Bench "pasture") to a point where the drivers had stopped after the road climbed out of Hells Canyon, now inside the Red Rock "pasture." The entire court had just witnessed the fenceless area between two "pastures" that the DNM had used to create trespass charges on Tim's cattle in 1993!

They continued on a short distance to a little road that led out to the rim of the canyon overlooking Castle Park, where Charley's brother, Joe Mantle, was buried in 1950. Ever since, the family has referred to it as "Joe's grave road." Judge Kane declined an offer to drive across the road to the grave site. This was the area the TRO claimed as illegal off-road use with motorized vehicles. DNM had posted a sign "Closed—reseeded area." At this point the trip ended!

Closing Argument by Stanley F. Johnson before Judge John Kane

November 18, 1996
Mantle Ranches, Inc. VS, U.S. Park Service, et al
May it please the court:
Plaintiff asks the court to deny Defendants' motion for preliminary injunction. [TRO had been converted]

First: In each instance where Defendants have complained, administrative remedies are available – and have been for the many years that these conditions are alleged. In any event there has been no showing of immediate and irreparable harm – and there is no valid reason for the Defendants' motion.

Defendants' motion raises 5 issues: (1) alleged illegal dumping by the Mantles, (2) water development, (3) overgrazing, (4) road development and (5) off-road vehicle use.

Concerning the alleged illegal dumping: Here is a natural storage area—as the court may have observed during your recent inspection of the site—located several feet from the river and used for storage by the Mantles on land they thought was their own—at least until the survey just completed—since before the 1940's—along the river where 5,000 people traveled last season—apparently discovered coincidentally by Defendants *after* the Mantles filed their complaint in this court in 1995. Mr. Petersburg testified that he discovered a pile of magazines with the Mantle name at another place in the 1970's.

The Mantle storage area was discovered some 50 years after its use—in spite of surveillance, or monitoring, of the Mantle family and their operation by the Park Service during Mr. Huffman's tenure.

There is no evidence of hazardous materials and no showing of irreparable harm or injury to the Park Service.

Concerning the illegal water development: Your Honor, we submit that this is the one that blows the cork right out of the bottle!

Several words can be used to describe the conduct of the Park Service in the destruction of the Mantle spring boxes and pipe at Springs Tom 1 and 2. *Negligent? Precipitous? Outrageous? Sheer Arrogance? Inappropriate self help during the pendency of this civil action, without notice of intent to destroy?* CRIMINAL—in this regard we submit CRS 37-84-121 and CRS 37089-101, pertaining to penalties for refusal to deliver water and breaking gates or flumes, copies of which are here available. Your Honor, there are no mitigating circumstances here. We ask that a special prosecutor be appointed to determine whether Federal Law has been violated, whether an administrative inquiry should be ordered, and to cooperate with the district attorney of Moffat County in any investigation. This was total destruction of this diversion: Spring boxes broken, pipe pulled from the rock, and pipe filled with bentonite, and rags, and dumped on the ground. I submit, your honor, that they destroyed a part of our western heritage. These were structures of historical significance—constructed by a settler almost a hundred years ago. True, not exactly the same, perhaps, as the Roman aqueducts, Tom Blevins created the diversion—and maintained by the Mantles, was an example of this man's effort to survive in rugged terrain, in an awesome and unforgiving land.

Not only adjudicated by the Mantles, which in Colorado is confirmation of appropriation and grants a date in the priority system, the Mantles also claim the historical right to divert the spring water acquired from Tom Blevins: Mrs. Burdick remembers the diversion structure since 1912; Mrs. Queeda Walker remembers it from 1938;

Superintendent Lewis recommended the use of pipe to allow mobility as part of grazing management. We invite your attention to the Lode Law of 1866: (43 U.S.C. 932—Rights of Way—and 43 U.S.C. 661—concerning water).

In addition to the appointment of a special prosecutor to determine if any crime was committed by the Park Service personnel, we ask that the court order the Park Service to replace the boxes and pipe in the condition before they were destroyed—including the lead pipe in the rock.

Concerning the alleged overgrazing: Your Honor, there is no evidence before the court of resource damage caused by overgrazing—and the Mantles have been grazing cattle on this land for almost 80 years. There is no evidence that the Mantles were charged with violating their grazing permit in 1996.

Mr. Philippi—whose credentials are impressive in range management—made a study of the Mantle allotment over thirteen years and found no change since the 1971 soil conservation survey and concluded: That the Mantle range trend and condition are stable. There is no degradation from livestock grazing. The elk population impact is significant and not monitored. Water sources should be developed for livestock and wildlife distribution. There is no documentation of overgrazing. There should be an allotment management plan. The BLM supervised the Mantle grazing permits until 1961. The Park Service has been in charge since then, and the Mantle allotment has been reduced from 2,000 AUMs to the present 954 AUMs. With proper water development, Mr. Philippi testified the allotment could support 2,000 AUMs. [AUM means animal units per month authorized to graze an allotment. One cow or horse is a unit; one cow and her young calf is a unit.]

We submit that there has been no government testimony to the contrary. The Park Service complaint is confined to 15 acres of Laddie Park, according to the testimony of the government botanist. She counted 15-20 head of cattle there. As the court may have observed during your inspection at that site, the river is shallow enough for livestock and wildlife to cross the river almost all year round. But, as Mr. Huffman testified, about 5,000 people also used that site last season. This has historically been grazed as a part of the Mantle allotment.

With regard to alleged illegal road development: We submit that there has been no evidence of such development brought to this court. Defendants have admitted that the Mantles have historic access rights. These rights should include access to Joe's grave site. This is an easement by necessity. Your Honor, we submit that the Park Service's recent interest in seeding the road is a subterfuge to harass the Mantles. Mrs. Walker used the road to the grave site as early as 1946. Lonnie Mantle also testified that he and others used this road to park vehicles while attending school, hauling wood in the 1940s. Evidence has shown that road graders were used since the 1950s on Red Rock Road and on the Yampa Bench Road—County Road #14—since the early 1940s. They should continue to do so.

As Tim Mantle testified, the Mantles have operated under the 1975 Special Use Permit and there is no evidence that they have abused this permit. We submit that they should have access rights to maintain the springs and use vehicles for that purpose.

In Conclusion: Your Honor, this whole case concerns the issue of survival of the Mantles as a working ranch.

We submit that there is no merit in the Defendants' motion for preliminary injunction and ask that it be denied.

We ask that the Defendant be ordered to replace the boxes and pipes in Springs Tom 1 and 2.

We ask that a special prosecutor be named to investigate and pursue any criminal action concerning the destruction of the diversion structures at Springs Tom 1 and 2.

We ask that maintenance of Range improvements contained in the 1975 special use permit be ordered to continue until this case is resolved on its merits and the Park Service should be ordered to recognize the vested rights of the Mantles historically and by law.

A point of personal privilege, Your Honor. I have represented the Mantle family during a span of 37 years. When I was younger—much younger—I had the occasion to help in the spring roundup—riding from before dawn to late night, branding and servicing calves. I have attended most of their weddings and funerals and have seen them grow into four generations. With this in mind, I want you to know on their behalf and the attorneys here how much we have appreciated your patience, your interest, and the time you have spent, and your obvious dedication to the cause of justice in this case.

Thank You!

Judge Kane rendered his MEMORANDUM OPINION AND ORDER on November 18, 1996.

1. The dump is not a dump—it is a storage area. Mantles were ordered to clean up offending objects and restore it to its historical condition as a hay yard.

2. Illegal water development: The NPS unlawfully removed the spring box and pipes leading from it. The NPS was ordered to restore the spring box and pipes to their former condition at their own expense.

3. Excessive grazing: campsites and livestock mixed warranted no overgrazing. Motion denied.

4. Illegal road and off-road use: No evidence of road construction or improvement. Historic access to Joe's grave restored as "an easement of necessity."

Judge Kane ruled in favor of the Mantles on all charges but one: They had thirty days to clean up the storage site and restore it to its original hay yard capacity. The NPS immediately appealed the judge's ruling on all counts. On January 13, 1997, *Judge Kane stood by his decision*, stating that if Congress seeks to "restore" DNM and the adjacent properties belonging to the Mantles to its original pristine state, it must do so by budgeting the necessary funds to condemn such properties, rather than taking the property rights by the process of "regulatory whittling"!

On the issue of off-road use, he also amended his order so that ATV and snowmobile use was allowed on the county road and other uses as directed by the 1975 Special Use Permits (SUPS). *He had just restored the 1975 Special Use Permits as the governing document on the Mantle allotment!*

The family's lawyers viewed this move by Judge Kane as "shots fired over their [Dinosaur National Monument's] bow." The National Park Service filed an appeal to a higher court but later dropped it. The old fox, Stan Johnson, was right: the TRO had been the trial. Now all that remained was picking up the pieces.

One month later, Huffman resigned as superintendent effective January 1997, and Steven Petersburg followed shortly after. News releases portrayed their retirements as a direct result of their unpopular stance against the Mantle Ranch.

The Mantle family, unlike many other ranchers who fought for survival, had no criminal convictions, and their greatest personal victory was that there is no ranger station in Castle Park.

Epilogue

AT LAST, THE FAMILY WAS FREE to conduct business at their ranch, with the United States Park Service held at leash by the mandates of a Federal judge. They could run cattle, run a tourist service of some kind, or they could sell the ranch.

Still, the "stink" of control and surveillance of their every move by government agents hung heavily over them. One thing was sure, it would never get any better because they lived in the middle of a national monument controlled by government regulations.

Charley and Evelyn Mantle had provided each of their five children with the means—both educational and financial—to have lives of their own. Adversity under the United States Park Service was something they did not relish. Their victory in United States Federal court had now set them free to make a decision about what they wanted to do with the Mantle Ranch.

The family offered the United States Park Service the opportunity to buy the ranch. But purchasing this beautiful jewel sitting in the midst of an existing national monument for the enjoyment of future generations was not to be. The Mantles made the offer, and the National Park Service flatly refused it.

The Mantles then sold their beloved ranch to a man who deserved to own such a treasure. He bought the ranch in 2004 and paid the appraised value of $6 million for it. The most the Park Service ever offered was $360,000. The new owner of the Mantle Ranch maintains the original ranch flavor with hay fields, a tended orchard, and cattle grazing on the range. Once again there is a vegetable garden, green lawn, and happy white hens scratching in the good earth. He has also made a few upgrades, like a big generator and mechanical farm machinery, which make life easier. He generously invited Mantles to come enjoy the Ranch any time they wish.

After Pat's death, the Pat Mantle Memorial Saddle Bronc Riding Championship event was set up in Steamboat. At the Steamboat rodeo each fall a separate competition is held between the six top saddle bronc riders of the year in amateur rodeos in Colorado. The winner is awarded $2,000 and a Winchester rifle engraved with Pat's name.

Pat's son, Steve Mantle, and his two sons and daughter live on Steve's ranch outside Wheatland, Wyoming. Steve works under a contract with the BLM wild horse relocation program. His awesome talent for handling horses is surely inherited through his Grandpa Charley and his father Pat. Steve and his two sons are in great demand to put on seminars about gentling and training unbroken horses around the country.

Tim, LaRue and their son Dean and his wife, Brooke and two grandsons, Bailey and Doak, live on their ranches near Rifle, Colorado.

Queeda and her husband Rex live in Boulder, Colorado. They have four children: Cindy and her husband, Dan Lisco; Freda and her husband, Mark Bishop; and Cody and his wife, Kathy now operate the Sombrero Ranches. Rex is still very active in the business side of the operation. [Justin and his family live in Tennessee.]

Chronological History of the Mantle Family

First Generation Mantles order of birth in Vernal, Utah

Father - John Mantle
Mother - Mary Ann Jolly Mantle
Charley Mantle - August 8, 1893
Nancy Mantle - 1895
Lewellyn [Wellen] Mantle - September 5, 1897. Died at 35 yrs. 9 mos. 18 days.
Joe Mantle
Bryan Mantle - raised by Hackings in Vernal
Lena Mantle - mother died in childbirth. Adopted at birth
1893 - August Charles (Charley) Thomas Mantle born in Vernal, Utah
1904 - Tom Blevins and Moran came to Pool Creek looking to settle.
Tom went on to Red Rock to homestead
Both got mail-order brides. Moran's bride unhappy and left right away.
Tom's bride stayed on awhile, but unhappily swept the dirt floor to bedrock.
1907 - February 18 - Evelyn Fuller Mantle born to Evan and Julia Fuller in New York.
1909 - January 14 - Wm. Franklin Fuller born to Evan and Julia Fuller
1910 - Chews bought Pool Creek from Moran
Jimmy Monahan was surveying to jump the claim. They beat him to it.

1915 - Dinosaur National Monument was created—eighty acres at a dinosaur quarry site located near Jensen, Utah
1917 - Charley in the army. Dec. 15, 1917 to February 12, 1919 Age at enlistment 24 yrs. 8 mo.
1918 - After discharge, Charley bought squatters rights to Mantle Ranch from Billy Hall
1926 - August 2 - Charley and Evelyn married
1927 - October 13 - Potch (Charles, Jr.) Mantle born
1928 - Wellen married Lorraine. Lived in Regina, New Mexico, since 1927. Disability began October 19, 1922
1929 - August 27 - Pat Mantle born
Nancy, Tab, and Herb spent summer at Mantle Ranch
Joe and Ruth Haslem married - came down to visit
Late fall Clark Feltch and Charley packed cottonseed cake from Serviceberry Gap on string of burros and broncs.
1930 - 1931 - Mantle family with cattle wintered on Ashley Creek at Uncle Hy Mantle's Ranch
1932 - Winter horse fell thru ice. Kids had a goat on sled
1932 - Summer severe drought – Family camped out at Hardin's Hole to save cattle
Pat three in August
Potch five in October
Preparing to trade the Mantle Ranch and move to the White Ranch in New Mexico
1933 - April Wellen shot and killed by Brian. Wellen 35+9 mo+18 days. Brian eighteen
Move to New Mexico is off
1933 - March 9 - Queeda Evelyn Mantle born at Chew Ranch at Pool Creek
1934 - Evelyn's letters begin to her cousin Eva in New York
1934 - Bad cricket year before July. Worst year of drought and depression aftermath.
1935 - July 1 - Nav Mantle [Lonnie] born in Bear Valley at old Smith place—Powers living there
Built schoolhouse at the ranch in late fall.
Had first school. Started in dugout with Ruth Haslem teacher.
1936 - Teacher backed out so frigid school year spent in Bare Valley
1936 - New house on north side of Castle Rock begun
1938 - Mantles and Chews built roads from east and west ends to their ranches.

1938 - Dinosaur Monument expanded by 200,000 acres into the canyon country. New roads made it easy for rangers to constantly watch and "manage" the Mantles
1938 - November 23 - Tim Mantle born at Mantle Ranch
1939 - Perry Mansfield camping trip
       First car—A Chevy pickup
1939-1940 - Archaeologists lived in basement of new house and excavated Mantle Cave
1940 - School election - Evelyn new secretary
1940 - East Road finished down Hells Canyon
1941 - November - Pat very sick in hospital
1941 - Moved into new house
1942-43 - War
1942 - Rationing
1943- Potch and Pat started high school in Craig
1944 - Scoggin killed
1944 - Eva came from New York to visit—brought daughter Barbara and Isa Coles
1944 - Christmas Katharine—the teacher— left and Evelyn began teaching
1945 - October - Evelyn teaching. Stanley and Eula Chew boarded—five kids, five grades
1945 - Chill Scoggin born in Boulder, Colorado
1946 - Potch left home
1046 - Bought new yellow Jeep
       With development of the rubber life raft Mantles' peaceful life was destroyed. River traffic during spring runoff was heavy.
1946 - Fall - Queeda started high school in Craig
1946 - Fall - Bench School started with Evelyn teaching Lonnie, Tim and Chew kids [Stanley, Eula, Carol, George D—too young but came some]
1947 - Pat graduated high school in Craig, Colorado
1947 - Fall Red Rock school started with Evelyn teaching Lonnie, Tim, and Chew kids
       Queeda sophomore at Wasatch Academy
1948 - Grasshopper invasion
       Morgans sold K-Ranch
       Chews moved to Vernal, Utah
       Echo Park Dam became a real threat
       House school started with Evelyn teaching Lonnie and Tim
       Becky Morgan married - Queeda bridesmaid

CU excavated Hells Midden [Bob Lister]
1949 - Eva came out West with daughter Barbara and husband Billie Reymore [married year before]
Joe died and was buried on the Mantle Ranch he loved
Fall - Lonnie started high school at Wasatch Academy
Thanksgiving - Bought new Ford truck. Visited Lonnie and Queeda at Wasatch
1949 - 1950 - Fall - Potch worked for Rial Chew
1949 - 1950 - Pat worked for Calder at K-Ranch
1950 - Queeda graduated high school
Doctor told Evelyn she had a heart condition
Julia Fuller is bedridden.
Queeda rodeo queen
1950 - 1951- Last year of school at the ranch
1951 - Tim started eighth grade at Wasatch Academy
May - all coal mines shut down in Oak Creek
Fall - Charley out to doctor in Ogden for his back
Winter - Charley planning trip to Mexico
Winter - Pat managing ranch - would be getting drafted soon
Christmas - Family nearly froze to death getting home. Charley pneumonia. Rescued by Dick Toole and Moffat County road crew
1952 - Charley and Evelyn and Pierre and Hilda Morgan trip to Mexico. Pat at Ranch
Devastating winter—thousands of livestock died.
1952 - Pat drafted into army
1953 - Lonnie graduated high school
Dick Toole December broke down at Thanksgiving Draw and nearly perished. He and wife Maude walked sixteen hours to get to the Mantle house. He brought in a trailer house for winter camp so nobody would ever suffer such danger again. Lonnie and Tim start a deer hunting camp with Portuguese hunters.
1953 - Lonnie started college at Colorado A & M
1954 - Queeda graduated college. Christmas in Kansas. Worked CU History Department then later at Shell Oil Co in Denver.
1955 - Tim 16 - senior at Wasatch
Lonnie sophomore at CSU
Pat at home
Julia Fuller died mid-winter
Charley Mantle, Jr. [Potch] married to Marmion Hart in Oregon

Queeda and Rex Walker married and moved to Caracas to work for oil company
1956 - Tim graduated high school
Tim started college at CSU
Steve Mantle born December 24. First grandchild of Charley and Evelyn
1957 - Lonnie graduated college
Evelyn went to Tyler, Texas, for the birth of Queeda's baby
Cindy Evelyn Walker born in Tyler, Texas, December 12
Cammie Lynn Mantle born December 24 in Dickinson, North Dakota.
1958 - January - Queeda and Rex moved to Colorado to a rental house on Cherryvale
1958 - February - Lonnie drafted - 6 months in Fort Hood, then Germany
1959 - Sombrero Ranches started. Pat was living at Gadds
1959 - Rex and Queeda bought Cherryvale farm
1959 - August 9 - Justin Ross Walker born
1960 - February - Lonnie discharged
1960 - Tim graduated college
1960-1961 - Tim drafted into army
1960 - July-March 1961 - Charley travels Brasil ranch country from west to east and in between.
1960 - Dinosaur Monument boundaries expanded - took in all of Mantle Ranch holdings.
Pat started Sombrero Stables in Grand Lake
1961 - Lonnie, Tim, Grace rodeoed local rodeos.
1962 - January - Charley and Potch went to Australia
1962 - Lonnie Mantle and Grace Spykstra married January 20
Lonnie and Grace moved to Wyoming
1962 - Chari Loy Mantle born on May 19 in Dickinson, North Dakota. Evelyn there.
1963 - Kail Everet Mantle born on May 4 in Riverton, Wyoming
1964 - Tim Mantle and LaRue Betke were married in Boulder
1963 and 1964 - Severe winters wiped out Potch and Marmion's cattle herd.
1964 - Summer - Charley severe gallstone attack in Halliday, North Dakota, and to avoid surgery, retreated to Glendive, Montana, to Jack Eaton's ranch
1965 - Freda Kay Walker born February 12 in Boulder, Colorado
1965 - Dean Mantle born

1965 - May - Charley filed application with NPS to amend his patent which incorrectly described boundaries of the Mantle Ranch homestead.
1965 - Spring - Potch and family moved back to Glendive, Montana
1965 - Summer - Charley signed the Mantle Ranch over to his five children
1965 - Summer - Charley spent time at Sombrero Stables in Estes Park, then Boulder
1965 - Summer - Charley bought new gray Willys Jeep in Longmont
1965 - Charley drove to Montana, returned with horse [Montana] for Cammie
1965 - Evelyn left the ranch for winters
1965 - Lonnie first shoeing chute
1965 - December 1 - Mickie Mantle born
1965 - Charley spent December through February in Sinaloa, Mexico
1965 - December through January 1966 - Evelyn in New York
1966 - Marmion and girls move to Colorado and managed KOA Campground in Estes Park
1966 - Queeda running Sombrero Stables in Estes Park.
Pat producing rodeos all over Western Slope
Rex producing Little Britches Rodeos on Eastern Slope.
1966 - Lonnie got his first dude horse contract
1966 - Tim dug a well at summer camp under old cabin. At twenty feet struck a good flow
Also built adjoining reservoirs in long draw to collect drainage.
1966 - Artesia renamed Dinosaur
1966 - Summer - Evelyn at Mantle Ranch - Tim piped water to campground and house
1966 - Summer - Rex sent Evelyn a German Shepherd pup (Zoro)
1966 - Queeda visited the ranch with three kids and Marmion with two kids.
Four oldest stayed awhile
Queeda and Marmion cannot stay as both working
1966 - Summer - Charley in Wyoming with Lonnie
1966 - August - Charley went to work for Jack Eaton
1966 - Evelyn began working for Fashion Bar as seamstress. Stayed six years, then six more years as private seamstress in Boulder.
1966-67 - Marmion in Boulder – Cammie eight years, Chari three years

1966 - August - Tim and Boyd Walker in Montana to visit Charley. Big Douglas deal.

1966 - October - Charley in Glendive at Jack's. Applied for SS at 72. Needs birth certificate to prove age. Sent QW his passport, needs his army discharge. It is somewhere in a safety deposit box. Found it!

1967 - January 1 - Charley worked as a cowboy for Jack Eaton at Foss Place.
Letter February 12 worried about Mickie
Letter July 30 to family in Estes Park Sombrero. I worry all the time about little Mickie. I didn't get my wellfare check for June

1967 - May 20 - Darlene Mantle born in Riverton, Wyoming

1967 - Marmion and Evelyn moving to a trailer house on Darryl Lemons farm

1967 - Rex Walker bought High Country Stables in Rocky Mountain National Park

1967 - Billie Reymore and Aunt Harriet died

1967 - Sombrero bought Craig Ranch at Big Gulch -7,000 acres

1967 - Pat moved to Craig to take care of Sombrero Ranches Western Slope

1968 - Bought Brown Park Ranch, Boon Draw, and Thompson Basin.
15,000 deeded acres plus other leases and BLM made 110,000 acres of pasture

1968 - Sombrero horse drive began from Craig ranch to Browns Park

1968 - Charley spent year from January to fall at Red Rock. Looked at an abandoned schoolhouse he wanted to move onto the place for his home.

1968 - Rex spent most of February in Roosevelt, Utah, at an oil well. Dry!

1968 - Walker family and children and friends ski a lot.

1968 - February - Grace and kids down. Mickey getting allergy medicine series

1968 - Keith and Eileen Hagler and her three kids working for Sombrero Ranches. Moved into Cherryvale house in fall.

1968 - Rex and Queeda sold Cherryvale farm

1968 - Spring break - Marmie and Queeda took kids and visited Grandpa at the Mantle Ranch.

1968 - Mike Harding managing Grand Lake

1968 - September - Queeda moved into new house on Jay Road

1968 - September 28 - Cody Rex Walker born
1968 - Cammie is Boulder Pow Wow Little Britches Princess
1968 - Cammie very proud of braided reins Grandpa made for her.
1968 - Rex's sister, Wanda, and family moved back to Tyler from Mississippi.
1968 - Pat started Sombrero Steamboat Stable and also put on a weekend rodeo
1968 - December - Charley takes new jeep and heads for Mexico for the winter.
1968 - Rosa Jurado in Boulder for school year.
1969 - Evelyn and Marmion move into new house on Jay Road
1969 - January 5 - Charley killed in Mexico. Funeral in Craig. Buried on Red Rock
Age: 76+4 mo+27 days
Tim went to Arizona to pick up the Jeep
Lonnie picked up Charley's casket in Denver and hauled it to funeral home in Craig
1970 - Potch died. Buried December on Red Rock. Age 42
1970 - Rex Walker bought Horse Country Club, a boarding facility in Scottsdale, Arizona
1970 - Marmion working at KOA again
1971 - A killer winter on Western Slope. Earl Vaughn to help Pat. Mike to help Tim
1971 - Cindy in Mexico for school year.
1971 - Tough winter. So much snow in January. Tim, Mike, Pat all using snowmobiles. Pat got sued for starving horses. Case thrown out of court.
Tim and LaRue leased out snowmobile business and moved to Greystone to look after their livestock
1971 - Tim living at Greystone at Bower Place
1971 - February - Tim and LaRue had stillborn baby girl. Buried at Red Rock
1971 - March - Aunt Nancy at Queeda house. Terribly crippled
1971 - January 29 - Rex, Queeda, Freda, Cody in Scottsdale. Drove a pickup pulling a trailer. Justin stayed with Granger. Cindy in Mexico
1973 - July - Cody Walker kicked by a horse and nearly killed
1974 - Evelyn sewing for private customers
1974 - Wilderness status proposed for Dinosaur National Monument

1974 - Lonnie, Grace, and grandchildren to Evelyn's house for Christmas
Walkers away in Texas for Christmas
1983 - Dean Mantle wrestled for Meeker Cowboys in high school at 119#
1992 - March - Pat Mantle died and was buried on the Ranch in Browns Park
1994 - November 21 - Mantles filed a lawsuit in Federal Court against the National Park Service
2001 - October 16 - Keith Hagler funeral
2004 - Mantle Ranch sold
2005 - June 15 - Sue Mantle died
2007 - July 25 - Grace Mantle died in Longmont, Colorado, at Longmont Community Hospital. Drs. Al and Kay Carr in attendance along with all her family.

Sequence of patent revision went like this:
1965 - Request sent in by Charley
1974 - April - Land conveyed to Mantles
1974 - July 8 - Conveyance denied
1977 - March 15 - paperwork resubmitted to BLM
1977 - Patent revisions on Mantle Ranch finally made. Restored two 40-acre pieces of bottomland exchanging the 80 acres of sheer cliffs of no use called No-Mans-Land.
1978 - By October still no reply. Zeke Scher of *Denver Post* "Inquired"
1978 - This "nudge" caused it to be done and approved but not before Evelyn died
1978 - January 2 - Evelyn died and was buried in Boulder